DOCUMENT REPRODUCTION IN LIBRARIES

Document Reproduction in Libraries

by

DONALD MASON

Fellow of the Library Association
Affiliate Member of the National Microfilm Association
LIBRARIAN, IMPERIAL CHEMICAL INDUSTRIES LIMITED,
DYESTUFFS DIVISION

LONDON
ASSOCIATION OF ASSISTANT LIBRARIANS
Group of the Library Association
1968

First published 1968

SBN 900092 01 7

TO
JENNIFER AND ROBERT
JOHN AND JEFFREY

Printed in Great Britain at the St Ann's Press
Park Road, Altrincham

Preface

This book endeavours to present to students of librarianship an outline of the major aspects of document reproduction services in libraries, and restricts itself in the main to those processes which are available commercially. It does not set out to be an operating manual for document reproduction units because such a work would require a mass of technical detail that would make it unsuitable for students. Nevertheless it is hoped that it will be of assistance to practising librarians, particularly in the selection of equipment.

Chapters 2 and 3 are expansions of those chapters on the same subjects which appeared in my *Primer of non-book materials in libraries* published by the Association of Assistant Librarians in 1958.

D. MASON

Torkington, Cheshire
April 1967

Contents

Figures in the text

Plates (between pages 80 and 81)

ACKNOWLEDGEMENTS

Photographs are reproduced by kind permission of the following manufacturers: Roneo Ltd., Roneo House, Lansdowne Road, Croydon (Plate 1); Gestetner Duplicators Ltd., Gestetner House, 210 Euston Road, London N.W.1 (Plate 2); Rotaprint Ltd., Rotaprint House, Honeypot Lane, London N.W.9 (Plate 4); Varityper, Maylands Avenue, Hemel Hempstead (Plates 3, 5); Rank Xerox Ltd., Rank Xerox House, 338 Euston Road, London N.W.1 (Plate 6); Kodak Ltd., Recordak Division, Kodak House, Kingsway, London W.C.2 (Plates 7, 8, 9, 10); Caps Equipment Ltd., 7 Westmoreland Road, London N.W.9 (Plate 11).

CHAPTER I

Office Printing and Duplicating

The increasing amount of printed material which is constantly being published, and which has to be catalogued, indexed, stored, and retrieved on demand, gives rise to many a groan from librarians contemplating the work load. This does not deter most of us from adding to the pile with our own brain children. However we can console ourselves with the knowledge that *our* publications are the guides by which readers can find their way into the larger mass of material which awaits them. For printing and/or duplicating the average library requires moderately priced equipment that will produce prints on octavo, quarto and foolscap paper but probably not larger. (If the new International sizes are being used then the paper sizes required are A4, A5 and possibly A6.) If the machine can also be used for printing 5 in. × 3 in. cards so much the better. Because the printed material must be read easily, the quality of printing must be fairly high. If possible, the readability must be improved by use of headlines and different founts. The print must also be of a permanent nature (the thought of catalogue entries fading into oblivion will, I trust, strike terror into the stoutest heart). One other major point to bear in mind is that when a particular item has to be printed more than once in different forms (e.g. details of a book printed on a catalogue card and in an accessions bulletin), then the process should if possible be so arranged that the typing is done only once and the resultant matter used in all the different forms.

The cost of elaborate and complicated machinery is too high to justify purchase for occasional use. Equipment of this type must produce a large output in order to earn its keep. Fortunately there is a wide range of reasonably priced machinery which will serve the purposes of most libraries. Catalogue cards, reading lists and bulletins can be satisfactorily produced on stencil duplicators. If offset-litho has been installed to print various forms, tickets, etc., then this can be used also for the other publications and does give a rather better print. The day-to-day needs of the service should be analysed and the most suitable equipment purchased. The special one-off jobs which occur every now and then can be put out in part or entirety to printing and duplicating agencies.

Dye Transfer Process

Sometimes known as the Hectographic Process, the Dye Transfer process was the first system to be developed for the copying of letters and other office papers. The invention is credited to James Watt, although in its original form it did not prove practicable. Towards the end of the 19th century an improved process using gelatine was developed and this was further improved in the 1930's with the introduction of the spirit method. The basic principle is the same in both methods. The matter to be copied has to be typed or drawn on to a master using some material which contains an aniline dye. The dye can be held in special typewriter ribbons, carbon papers, inks and pencils. The matter is then transferred from the master to sheets of paper, copy by copy, until the dye is used up and the copies become too weak to be legible. By typing or drawing various parts of the image in different coloured dyes, a multi-coloured copy can be printed at one go.

In the gelatine process the typing and/or drawing is done on the face of the master. This is then placed face downwards on a moistened gelatine surface. The dye transfers from the master to the gelatine where it forms a reversed image. A sheet of paper is then placed over the image, pressure is applied and the image transfers to the sheet where, of course, it is now the right way round again. The gelatine is usually in the form of a thin layer with a cloth backing to act as a support. It can be used in a flat bed machine or fastened round rollers in a rotary machine. It cannot be re-inked and so is only suitable for short runs up to approximately 100 copies.

In the spirit process the master consists of a sheet of hard surfaced paper, together with a sheet of dye impregnated carbon paper. For convenience master sets can be purchased which consist of the master sheet, the carbon paper and a backing sheet to protect the carbon all made up into one unit. The matter to be duplicated is typed and/or drawn on the face of the master, and this action takes up the dye from the carbon paper and deposits it on the back of the master in the form of a reversed image. The master is then separated from the carbon and the backing paper, and is placed on the cylinder in the rotary duplicating machine. Sheets of copy paper are fed through the machine and are first dampened with liquid solvent (hence the term 'spirit') then pressed against the master where the solvent permits the transfer of the dye from the master to the copy paper. Each turn of the cylinder produces one copy, and the sheets are automatically fed in, printed and then stacked. The solvent evaporates almost at once so there is no smearing of the copy nor offset from one copy to the back of the copy stacked above it. Spirit duplicators are available from several companies including Block and Anderson Ltd. (*Banda*), Ditto (Britain) Ltd., Dupl-o-Print Co., Fordigraph Ltd., and Office Machinery Ltd. (*Copyrex*).

In the process described above care must be taken to avoid handling the carbon paper otherwise the operator's fingers can become stained with the dye and this can in turn mark the copies. Another method (*Azograph*, *Chemograph*, *Bandagraph*) has been developed to avoid this problem. The master is prepared as for normal spirit duplication except that a transfer sheet replaces the carbon sheet. The transfer sheet is coated with a material which contains chemicals that will combine with the solvent to form a dye. Thus all the chemical action takes place within the duplicating machine and both the master and the transfer sheet can be handled by the operator without fear of staining. *Azograph* (A. B. Dick Co.) and *Bandagraph* (Block and Anderson Ltd.) have two colour forming compounds on the transfer sheet that are held separated and therefore colourless until they are brought together and developed by the action of the solvent in the duplicating machine. The *Chemograph* (Ditto (Britain) Ltd.) is similar to these, but has one of the compounds on the transfer sheet and the other as a transparent coating on the copy paper. This means that only this special copy paper can be used with the *Chemograph* process. A similar method which requires special copy paper is the *Merlin* dry process (Ellams Ltd.). In this development is produced entirely by the reaction between the master bearing its carbon image and the coated paper. No spirit is used in the duplicating machine, the process being completely dry.

Masters for the spirit duplication process can be made by the thermographic process (*Thermofax*, Eichner *Dry-Copy*). (See pages 47–9.) The original material to be copied is passed through the copying machine together with a master sheet and a special sheet which contains the dye held in a wax. The heat melts the wax and the image is transferred to the master sheet which can then be used to run off the required number of copies. Masters can also be prepared by Xerography (see pages 49–52). The *Xerox Flo-Set* unit is used in conjunction with the *Xerox Model* 1385 for this purpose. The advantages of producing masters by either of the methods are (a) the handling of carbon paper is avoided, (b) a master can be prepared from an existing original without the need to re-type (c) drawings can be put easily on to the master.

Stencil process

The principle of stencil duplicating is based on the passing of ink through openings in the stencil and on to the copy paper. The stencil master is usually made up in three parts, the stencil itself which consists of a thin sheet of paper with a coating which is impervious to ink, a backing sheet, and between these two a sheet of carbon paper. The matter to be copied is typed or drawn on to the stencil. Most typewriters have a stencil setting which suppresses the ribbon and allows the metal keys to strike the stencil directly. When the stencil is struck by the key, the protective coating is removed and an opening is created in the form of the type character on the key. At the same time a copy is produced via the carbon sheet on to the backing sheet. The primary use of the backing sheet is to absorb the blows of the typewriter keys, and its secondary use is to act as a surface for the carbon copy in order to facilitate proof-reading. Errors on the stencil can be corrected by the application of a fluid which restores the impervious surface and then by cutting the correct letters.

When the stencil has been completed, it is separated from the backing sheet and placed in position on the drum in a duplicating machine. (See plate 1.) The drum consists of a perforated metal cylinder with a cloth covering, and has a supply of ink inside it. The drum is turned either by hand or by an electric motor, and with each rotation a piece of copy paper is fed through between the drum and a pressure roller. The ink is squeezed through the perforations in the stencil and is deposited on the paper. The paper then moves on to a delivery tray where it is stacked ready for use,

Rotary stencil duplicating machines of varying degrees of complexity are made by many companies (e.g. Roneo, Remington-Rand, Gestetner). Some have two drums which are interconnected with a silk-screen on to which the stencil is affixed. Mention must also be made of flat-bed duplicators. In these each piece of copy paper is placed under a framed silk screen to which the stencil is attached. An inked roller is passed across the face of the stencil and ink is passed down through the perforations on to the copy paper. This, as the description shows, is a somewhat laborious process.

To avoid offset from a copy to the one stacked above it, it is customary to use paper which is slightly absorbent and thus soaks up the ink quickly and retains it. If hard-surfaced papers or cards are to be used it is often necessary either to interleave the copies with absorbent paper or to spread each copy out to dry as soon as it is printed. This problem is being overcome by the development of better inks; also a device to introduce a barrier between each copy has been brought out by Roneo. It is known as the *Electro-spray* and when fitted to a Roneo duplicating machine it sprays a small quantity of powder on to each copy as it emerges from the machine. The powder physically separates the copies and allows the ink to dry.

Simple drawings and lettering can be put on to stencils with the aid of stylus pens. The stencil must have a hard-surfaced backing plate placed behind it, then the drawing is made with the metal point of the stylus pen. A variety of effects can be obtained with different pens, e.g. a loop pen gives a thick and thin stroke like a nib, a ball-point stylus pen allows a free action in the drawing, and wheel pens give dotted lines of several types. Lettering can be done by using a plastic lettering guide. The guide is placed on the stencil and a stylus pen is drawn round the outlines of the required letters. Background shading in various textures can be added to the stencil by the use of a shading plate. This is a plastic sheet with the pattern embossed on its surface. It is placed underneath the stencil and then a burnisher is rubbed across the selected areas. There is also an electrically-operated hand-held stencil cutting device known as the *Stencilograph*. It is held in the same way as a pencil, switches itself on and off automatically and will perforate a stencil at the rate of 3,000 strokes per minute. For posters and illustrations a brush stencil can be used. A brush loaded with a special solvent is used to draw designs and letters on the stencil. Where the solvent is applied the protective coating is dissolved, therefore every brush stroke is reproduced when the printing ink passes through the brushed areas on to the copy paper.

For complicated drawings or the reproduction of existing matter either the photographic stencil or the electronic stencil can be used. The photographic stencil has a sensitized surface which is exposed to light via the original. After exposure it is developed and the text on the original is reproduced on the stencil in the form of apertures. The equipment for developing the stencil consists of three chemical dishes designed to hold a small hand rotated drum. After exposure the stencil is fastened round the drum and then rotated through the chemical solution in each of the dishes in turn. After this it is dried and is then ready for printing.

Photographic stencils are available in two degrees of light-sensitivity. One can be handled in subdued room lighting and is used to make same-size copies from translucent, single sided originals. The other must be handled only is darkroom conditions, i.e. with a safelight, and is used via an optical device to make copies from opaque and/or doubled sided originals. A device for this process is made by Gestetner and is known as the *Photoscope*.

There are two models available, the first will only make same-size copies, whereas the second will enlarge or reduce. This model is basically a camera. The original to be copied is placed in the base of the machine and is illuminated by built-in lights. The image of the original is focused on to a screen at the top of the machine and the positions of the lens and of the copyholder are adjusted until the image is the required size. The lights are switched off and the focusing screen is taken out and replaced by a sensitized stencil which is held in place by a vacuum blanket. The exposure is made and then the stencil is removed and developed by the method already described. This method can be used for line drawings and half-tone prints, and is particularly useful where the duplication of columns of figures is required.

The electronic stencil is made on a facsimile copying machine (see plate 2). This machine has two cylinders, and the original material is fastened to one cylinder and the stencil to the other. A switch is turned on and the two cylinders commence to revolve. An electronic device scans the original material and activates an electric stylus which reproduces the original on to the stencil in the form of minute apertures. When completed the stencil is removed and then used in the normal manner. By this means copies can be made of printed materials, line drawings, half-tones and, with some machines, even photographs. The cost of a facsimile copying machine is such that few libraries are likely to have sufficient use to justify the expense. However in most large towns there are services which will make electronic stencils on demand at reasonable cost.

Offset Lithography

Lithography has developed from a process in which the matter to be printed was first drawn upon a stone surface using a greasy medium. The stone was then soaked with water, which kept the surface wet except for the areas occupied by the greasy medium. An ink roller was passed across the stone and the ink adhered to the medium but not to the wet part of the surface. Paper was then placed on the surface and pressure applied to make a print. The whole process depended upon the fact that water and grease do not mix. In modern office-printing the process has now been completely mechanized, (See plate 4.) A metal, paper or plastic

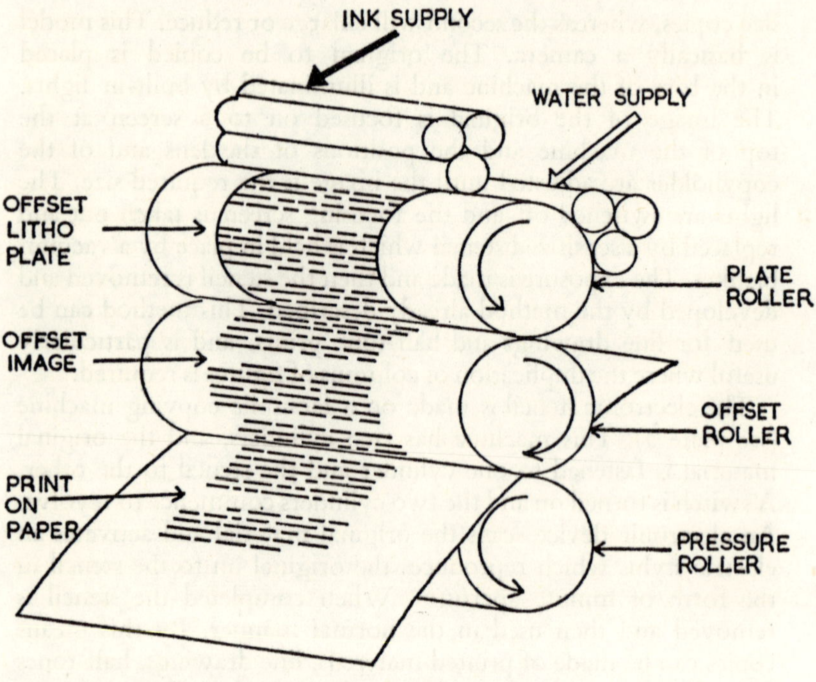

Figure 1. Offset Litho Printing.

Ink and water are fed on to rollers and are spread on to the offset litho plate. The plate roller turns and offsets the ink image on to a rubber blanket on the offset roller. This in turn transfers the image to the paper as it passes through.

plate is prepared by typing or drawing on the surface, a liquid repellent is applied to the plate and then it is fastened to the plate cylinder in the offset printer. (See fig. 1.) As the cylinder rotates ink and water are fed to the plate. The ink remains on the image but is prevented by the water from adhering to the non-image areas. The image is then offset to the rubber blanket cylinder where it is laterally reversed. The sheets of paper passing through the machine are pressed against the blanket cylinder by the impression cylinder, and the image is offset printed, this time the right way round, on to the paper. The use of the offset method thus allows the original matter to be typed on the face of the plate and thus be easily checked and errors corrected.

The quality and price of the plates varies from the cheap paper plates which can be used for short-run work of about 200 copies, to the metal plates which will give several thousand copies. Plates can be stored and used again as required. When the matter to be copied is typed on the plate, the typewriter must be fitted with a special ribbon, preferably a one-time ribbon. This consists of a greasy medium on an acetate base and, as the name implies, passes through the typewriter once only and is then discarded. Lines can be ruled on to the plate or drawings made with special inks, pencils or ball-point pens.

If the matter to be copied is already typed or printed, or includes existing drawings or half-tone prints, it can be put on to a plate by a photographic process. Pre-sensitized plates can be purchased which are usually of the diazo type so can be handled in subdued room lighting. A translucent master is made from the material to be copied, and then the master is used via the dyeline process to put the image on to the plate. In the *Azoflex* dyeline process, a carbon transfer foil is used to place the image on to an offset plate. The foil is exposed to light through the original and is then passed, together with the plate, through a developing bath. When the plate and the foil are brought into firm contact, the image, in the form of carbon, is transferred to the plate.

Other photocopying processes can be used to prepare plates from existing originals. In the mechanical transfer process (*Verifax*) the plate is used instead of ordinary paper and the image transferred directly from the matrix (see pages 41–3). Special transfer negatives are used in the chemical transfer process (see pages 39–41). The Kodak *Ektalith* mechanical transfer process will prepare an offset plate from an original in approximately two minutes and as a part of the equipment is a copy camera, the matter can be same size, reduced or enlarged. In the *Xerox* process the image can be placed straight on to the plate, and alterations in size are possible by using the copy camera (see pages 51–2).

Addressing machines

Apart from their original purpose, addressing machines can be used to produce printed copies when short runs are needed on several occasions (e.g. catalogue cards, periodical circulation slips), the plate being stored when not in use. The plate can be made of metal or plastic, or it can be in the form of a stencil or a spirit duplicating master.

The required matter is put on to metal plates in an embossing machine, which can be hand or power operated. It is not necessary, however, to purchase an embosser because finished plates can be made on a service basis by the manufacturers. Material put on to a plate can later be blanked out and replaced with other matter if so desired. The *Addressograph* plate is designed to fit into the lower part of a metal frame. The upper part of the frame holds a card strip which can be used as a heading for filing the plate. The *Bradma* is a single plate with strengthened edges. Both types of plate will carry up to nine lines of matter, and both can have metal tabs inserted on the upper edge to allow selective printing. Plastic plates are similar to metal plates but are more restricted in the amount of information that can be embossed on them.

The addressing machines range from small hand-operated machines in which each plate is inserted by hand and removed immediately after use, to large automatic machines which can print cards from plates at speeds up to 2000 an hour, and can print from all the plates or from selected plates only as required. The printing is effected by pressing the envelope or card against the plate with a carbon coated ribbon (like a wide typewriter ribbon) between them. The pressure is applied with a rubber pad, and if only a part of each plate is required to print, the rubber pad is cut to the shape of the part to be printed.

A typewriter is used to make both stencil and spirit masters, although for the former a special unit is attached to the typewriter to hold a stencil steady during typing. The spirit masters are produced in the form of a perforated roll, and after typing are separated into the individual masters. The stencil is supplied in a card frame, but the spirit masters have to be fastened into a card or plastic frame after typing. Both types can be stored for use at a later date. The printing process in each case has already been described in this book. The addressing machines for these two methods like those for the metal plates range from the hand operated to the power operated.

Equipment for the preparation of copy

In most libraries copy will be prepared on the ubiquitous typewriter, with occasional essays into art-work to provide such things as headlines and borders. There is however a considerable amount of equipment for producing good copy, ranging from expensive typewriters with changeable founts and built-in right hand margin justifiers, to simple and inexpensive sheets of letters and symbols for 'do-it-yourself' headlines, etc.

The *Varityper* is one of the best known variable fount machines (see plates 3,5). The type founts are in the semi-circular segments, which clip into a circular holder. Two segments can be held in place at any one time and are rapidly interchanged by rotating the holder. The number of characters printed per inch can be varied by a switch on the side of the machine, and the right hand margin can be justified. A whole range of founts is available for the machine, and also symbols, e.g. chemical and mathematical.

The *I.B.M. 72* electric typewriter uses an interchangeable spherical typing head. There are several different heads available and each one carries all the 88 keyboard characters in the appropriate fount. The heads are quickly and easily changed, the new head being simply clipped in to the mounting.

The typing of scientific and technical reports presents many problems not the least of which is the use of subscripts, superscripts and special symbols. To avoid the use of special typewriters for this purpose (or the writing in by hand of the symbols) a device has been invented which allows the insertion of one special character at a time into a holder next to the typewriter roller. The striking of a typewriter key then causes this special character to print. It can then be removed and either replaced with the next character required or ordinary typing can proceed. The special characters are known as *Typits*, and there is a very wide range of them available. The time taken to remove a *Typit* from its box, insert in the typewriter, print, and return to its box, is stated to be four seconds.

The *Justowriter* is an automatic tape operated composing machine with an electrical typewriter keyboard. It consists of two units. The Recorder unit is fitted with a punch which records on punched paper tape the code for each letter and function (e.g. Capitals, Carriage return, space) as the keyboard is operated. The copy is typed on this unit and at the end of each line a special Justifying Switch is depressed. This records the correct justifying information into the tape and the typist then proceeds to type the next line and so on. When the typing is completed the punched tape is transferred to the Reproducer Unit, where it will automatically produce a fully justified copy at a speed of 100 words per minute.

The *Marginator* is a device which can be fitted to almost any make of typewriter in general use. A draft copy is typed and then a special transparent scale is laid over the right hand margin. A satisfactory column width is selected visually and the indicator settings marked on the scale are noted. The *Marginator* is then set to the column width selected, and the final copy is typed with line by line adjustment of the *Marginator* which automatically condenses or extends the line to the desired width, thus justifying the right hand margin.

The *Flexowriter* is a punched tape typewriter that has been designed to cut out unnecessary repetition of typing effort. The copy is typed just once and at the same time a punched paper tape is produced with all the information coded onto it. Any further copies required are produced automatically by feeding the punched tape through a tape reader connected to the *Flexowriter*. This machine is being used by some libraries for the production of multiple sets of catalogue cards.

The *Optype* is an optical composing machine. The already typed copy is placed in the machine and each line in turn is visually examined by the operator through a sighting screen. One control is used to set interline spacing and another is used to set the right hand margin. The line is then seen exactly as it is required to appear in its final form. A further control is then used to photograph the line on to film. When all the lines on the copy have been thus treated the film is removed and developed. The resulting negative is normally used to print a positive image on to an offset plate. Apart from the controls for interline spacing and the justification, there are other devices which permit the operator to reduce, condense, bolden and/or italicise the type.

Another optical composing machine of a rather different type is the *Typro*. There is a choice of approximately 2000 different founts available on film for use with this machine, and it can carry up to 15 different founts for use on one photographic reel. The reel is inserted into the machine and a selector dial is rotated to bring the required character into the printing position. It appears in a magnifying viewer for a visual check, which also allows the operator to obtain the best spacing. The character is then photographically printed on to a roll of sensitized paper, which is held inside the machine. While the copy is being composed, the roll of paper can be fed through to a developer unit where it is developed, stabilised and dried. When the job is completed the roll is cut up into the required length of line and pasted up in page form. This paste-up is then used to make an offset plate by one of the many processes available.

The *Varityper Headliner* is a similar machine, the major difference being that each fount is supplied as a plastic disc about the size of a gramophone record, and the various characters are in negative form on the disc. The disc fits into the top of the machine, and the required character is obtained by rotating the disc with a control knob. The character is then printed onto a roll of paper and is processed automatically inside the machine. The finished print is ready in about a minute and a half. The *Headliner*, as its name implies, is principally designed for producing headlines and other display material for use with normal copy produced on the *Varityper*.

Most libraries do not have sufficient work to justify the purchase of headline composing machines. There are however inexpensive sheets of black letters in various founts and sizes from which 'do-it-yourself' headlines can be prepared. Two examples of these are (i) the *Artype* sheets, from which the self-adhesive letters are cut and then pressed firmly on to the master copy; and (ii) *Letraset* sheets from which the letters can be transferred to the master copy by the application of pressure (the easiest way to apply the pressure is by rubbing over the letter with a ball point pen).

Listing devices are of particular interest to those libraries which publish catalogues and reading lists. There are automatic list composers which would be ideal but for the fact that they are too expensive for only occasional use. They work on the principle of photographing sequentially a line or lines from the face of a series of cards. The cards are usually the same size as punched cards, and can indeed be punched and placed in correct order on a sorting machine. When the cards are in order they are fed into the listing machine and the information on the cards is photographed on to film. The film after development is used to make offset plates. Photographic paper can be used instead of film, and then used as a master to make copies on photocopying machines. Two examples of this type of listing machine are the *Varityper Foto-List* and the Friden *Compos-o-line*.

Lists can also be kept up-to-date and ready for immediate printing by using a strip index. Each entry is made onto a strip of thin card and is inserted into the index at its correct place. When printing is desired it is only necessary to check that each page of the strip index is balanced and has a page number. The pages are then used as masters to prepare offset plates or stencils by one of the many methods available. An example of this type of list is the Kalamazoo *Copy-strip*.

At the end of the office printing production line, the printed material has to be collated and bound. Most collation in libraries is done by hand, the large automatic collating machines being (once again) too expensive for occasional use. However there are smaller devices which can assist this operation. The Thomas *Gatherette* is an example of a portable electric collator which will hold up to six sets of paper sheets. When it is switched on the feed arms push the top sheet of each set forward towards the operator, who collects them, and stacks them at the side of the machine and then collects the next set. Collation can proceed at the rate of about 20 sets per minute, but is limited to not more than six sheets (i.e. twelve pages).

Collators of this type holding more than six sets of sheets are difficult to use because too many sheets issue for the operator to be able to grasp them easily.

Binding of material is usually accomplished in one of three ways:-

(i) Stapling; (ii) padding, using flexible glue; or (iii) plastic strip. Desk staplers will handle small unfolded publications, but are difficult to use when the staple is to be inserted into a fold. There are some desk staplers with extra long arms to reach to the middle of the paper, and there are also larger staplers, e.g. the British Brehmer Stapler, which have a special saddle attachment on which the folded publication can be rested while it is being stapled. When a publication is bound by having staples inserted down the left hand side, the staples should be covered by a gummed strip. This strengthens the spine of the publication, covers up the unsightly staples, and also acts as a protection from scratches.

The technique used in making tear-off paper pads can also be used to bind publications. The collated sheets complete with covers, if any, are stacked in a pile and flexible glue is coated on to the spines. To ensure good adhesion the edges must be fanned out or the backs scored with a sharp knive to allow penetration of the glue. Mull is then put over the glue and time allowed for the glue to set. Each publication is then separated from the stack by cutting along the mull. The spine can be protected by covering with a gummed strip. This method is not suitable for publications that will have heavy use, because with constant opening and closing the pages tend to work loose.

Plastic strip binding has the advantage of neat appearance and also allows the publication to be opened flat. The technique is as follows: first the pages have holes punched in them along the inner margin, the holes being approximately half-an-inch apart. A special punch is used which will take about 10 sheets at a time and makes all the holes at one punching. The sheets and covers are then taken to binding machines. The rolled plastic strip is inserted in the machine, where it is opened and the tongues on the strip inserted through the punched holes in the publication. The strip is then closed, thus locking the sheets together. Some machines have a control fitted which ensures that the plastic binding is set to the right diameter for the publication being bound. There are also some combined punching and binding machines, e.g. the General Binding Company's *Combo*.

CHAPTER 2

Photocopying

The chart opposite shows the relationships between the various methods of photocopying. There are, of course, other ways of grouping the different processes; for example, the first division could have been into camera and non-camera processes. This would, however, have grouped the *Photostat* with the micro-copying methods, whereas in its technique it is most nearly related to the non-camera processes. Although the term *photocopies* strictly applies to all forms of photographic copies, it has been used here in the narrower sense of copies which can be read with the naked eye. The term *microcopies* refers to those processes in which the copy has to be read with an enlarging optical device. The principles and uses of microcopies from a distinct discipline, and they have therefore been dealt with separately.

One of the major difficulties to the proper understanding of photocopying processes is the host of technical and trade names involved. Often there is no good alternative to the trade name for a process, and the trade name has become more or less a part of common speech. A well-known example of this is the term *Photostat*. This is a trade name, and its use to describe any other process is incorrect and actionable at law. Yet so many people will insist on calling any photocopy a Photostat. As far as possible trade names will be given in the description of each process.

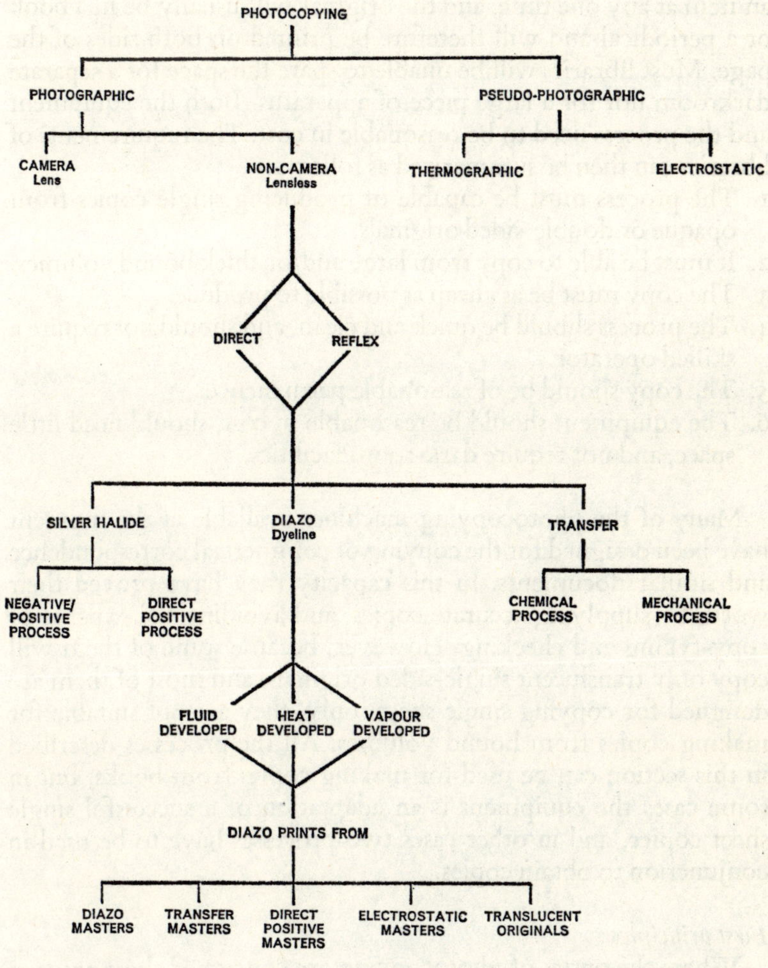

Figure 2. Photocopying chart.

Library requirements

Very few libraries are asked to produce more than one copy of an item at any one time, and the original will usually be in a book or a periodical and will therefore be printed on both sides of the page. Most libraries will be unable to spare the space for a separate darkroom nor for a large piece of apparatus. Both the equipment and the process need to be reasonable in cost. The requirements of libraries can then be summarized as follows:-

1. The process must be capable of producing single copies from opaque or double-sided originals.
2. It must be able to copy from large and/or thick bound volumes.
3. The copy must be as cheap as possible to produce.
4. The process should be quick and clean, and should not require a skilled operator.
5. The copy should be of reasonable permanence.
6. The equipment should be reasonable in cost, should need little space, and not require dark-room facilities.

Many of the photocopying machines available at the moment have been designed for the copying of commercial correspondence and similar documents. In this capacity they have proved their worth by supplying accurate copies, and avoiding the expense of copy-typing and checking. However, because some of them will copy only translucent single-sided originals, and most of them are designed for copying single sheets only, they are not suitable for making copies from bound volumes. All the processes described in this section can be used for making copies from books, but in some cases the equipment is an adaptation of a successful single sheet copier, and in other cases two processes have to be used in conjunction to obtain copies.

First principles

Where the optics of photocopying are concerned there are two basic methods. One uses a lens and the other relies on close contact between the original and the sensitized copying paper. When a lens is used the size of the copy can be varied, i.e. it can be an enlargement, the same size, or a reduction of the original. With the contact method the copy is always the same size as the original.

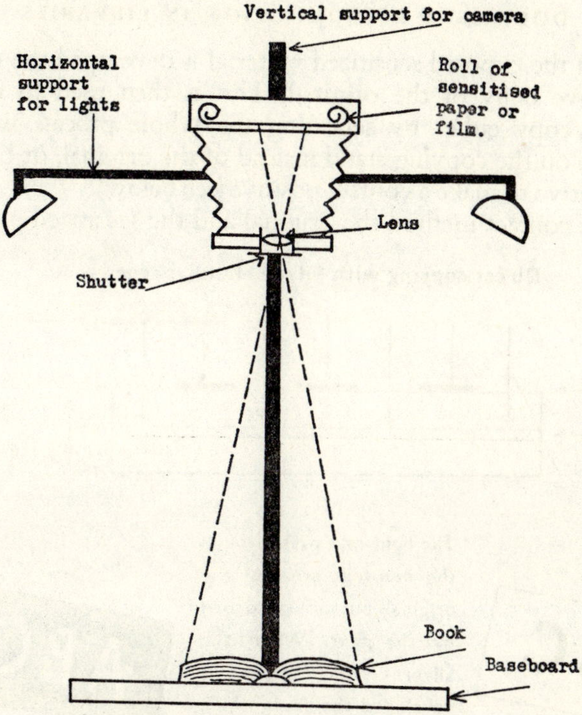

Figure 3. Copying with a camera.
The book is placed on the baseboard and illuminated. The shutter is opened for the correct exposure and a latent image is formed on the sensitized paper or film, which is later developed and fixed to form a permanent copy.

The method of copying using a lens is as follows:- the sensitized material (film or paper) is held on one side of a light tight compartment, i.e. a camera; the lens is situated on the opposite side and is covered by a shutter which prevents light from passing through the lens. Some device, usually bellows, is incorporated in the camera to allow the distance between the sensitized material and the lens to be varied. The document to be copied is placed opposite the lens and is illuminated. (See figure 3.) The shutter is removed and the reflected light from the surface of the document is transmitted through the lens to the sensitized material. The shutter is then closed and fresh sensitized material is placed in the back of the camera ready for the next exposure.

When the exposed sensitized material is developed the result is a negative copy of the original. This is then used to make a positive copy either by repeating the whole process with the negative on the copying stand instead of the original, or by using the negative to make a contact positive (see below).

In the contact method the original and the sensitized paper are

Direct copying with Silver Halide paper

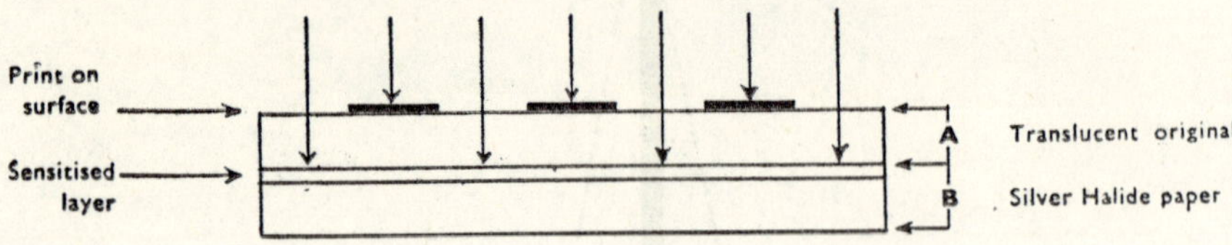

Original

The light-rays pass through the non-text areas of the original but are absorbed by the text. When the Silver Halide paper is developed the areas which have been exposed to light turn black. The result is a readable negative.

Readable Negative

Figure 4.

placed in the closest possible contact and then exposed to light. The apparatus used to make the exposure usually consists of a light box with a ground-glass screen and a lid which forms a pressure pad to hold the original and the paper in firm contact. Copies can be made in two ways. If the original is translucent and has text on one side only, it is placed facing the light source and the copying paper is placed with its sensitized surface against the back of the original. When the exposure is made the light passes through the white background of the original, but does not pass through the text because it is absorbed by the black print. The resulting copy is a readable negative and can be used instead of a positive. (See figure 4.) This is known as *direct copying*. It is also occasionally referred to as contact copying, but this is not an acceptable term because all non-camera methods rely on close contact between the original and the sensitized paper.

If the original is opaque or has text on both sides the *reflex copying* method has to be used. The copying paper is placed between the original and the light source, with the sensitized surface in

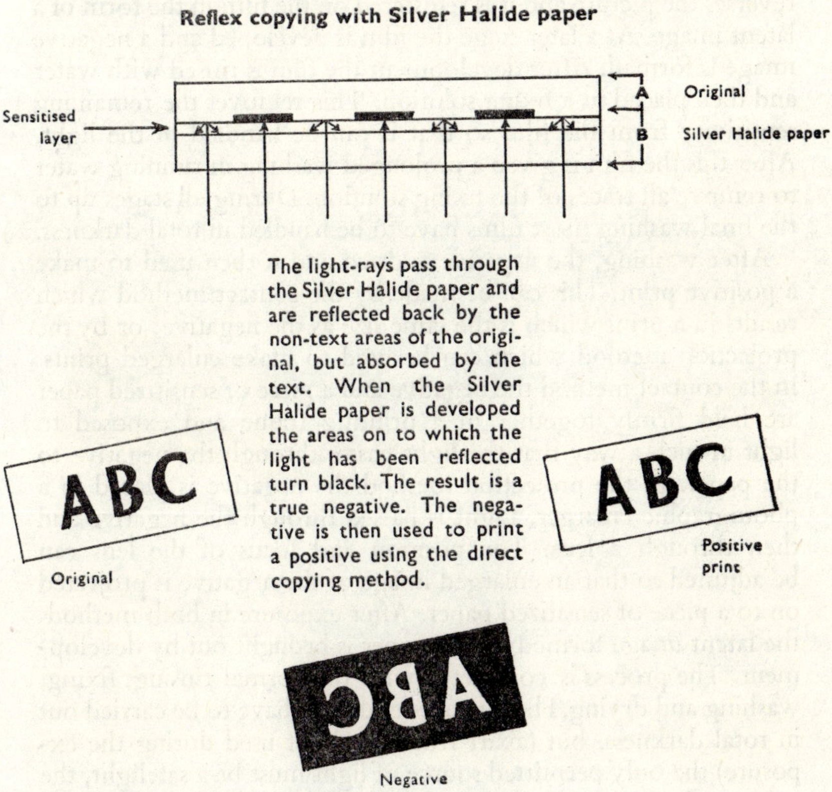

Figure 5.

close contact with the face of the original. When the exposure is made the light passes through the copying paper to the original. The black areas of text absorb the light, but the white background reflects light back to the sensitized surface of the copying paper. (See figure 5.) The resulting copy is a true negative. The direct copying method is used to make positive copies from negatives. The negatives can have been produced in a camera or by reflex copying.

Copying with a Camera

When a photograph is taken with an ordinary camera, the lens reverses the picture and it is registered on the film in the form of a latent image. At a later stage the film is developed and a negative image is formed. After development the film is rinsed with water and then placed in a fixing solution. This removes the remaining sensitivity from the film so that is can be handled in the light. After this the film is given a prolonged washing in running water to remove all traces of the fixing solution. During all stages up to the final washing most films have to be handled in total darkness.

After washing, the negative is dried and is then used to make a positive print. This can be made by the contact method which results in a print which is the same size as the negative; or by the projection method which can be used to make enlarged prints. In the contact method the negative and a piece of sensitized paper are held firmly together in a printing frame and exposed to light in such a way that the light passes through the negative to the paper. In the projection method the negative is placed in a photographic enlarger. Light is passed through the negative and then through a lens. The position and focus of the lens can be adjusted so that an enlarged image of the negative is projected on to a piece of sensitized paper. After exposure in both methods the latent image formed on the paper is brought out by development. The process is completed with the normal rinsing, fixing, washing and drying.These processes do not have to be carried out in total darkness, but (apart from the light used during the exposure) the only permitted source of light must be a safelight, the colour of which (usually orange) does not affect the sensitized paper.

Documents can be copied in this manner using a suitable camera. This must be of the type which has interchangeable lenses or which will take a supplementary lens for close up work. In order to avoid vibrations the camera has to be mounted on a stand and the exposure made using a wire cable release. Some form of balanced lighting needs to be provided in order to obtain good copies.

For normal document copying this method is too expensive. The lengthy process results in high labour costs. A darkroom is required and the necessity of working in subdued light, plus the use of chemical baths, make it unpleasant for anyone to spend long periods on this process. The modern photocopying processes are designed to avoid these defects as far as possible. Sensitized paper is used instead of film because it is cheaper, and for most copying purposes just as good. Dry or semi-dry development avoids soaking the copy in chemical baths, and most processes can be used in ordinary room light.

In some cases the use of camera and film for copying is to be preferred. When a coloured copy is required (e.g. when the original is an illuminated manuscript) the stand camera as described above will produce colour transparencies, either as positives or negatives. The positives can be used in a reader or projector, and from the negatives colour prints can be made.

Photocopying cameras

The best known apparatus in this category is the *Photostat.* This is essentially a large camera in which exposures are made on to a roll of sensitized paper. A prism in front of the lens laterally reverses the image, which means that the first copy is a readable negative. After exposure the paper is processed in covered dishes which are a part of the apparatus. The copy passes directly from the camera into the developing and fixing dishes thus allowing the *Photostat* to be operated in normal room lighting. When the copy is removed from the dishes it has to be washed and dried. There is another type of *Photostat* in which the processing is separate and can be done all together when the exposures are finished. In this case, however, the processing must be done in a dark-room or in a room with very subdued lighting.

The position of the lens between the sensitized paper and the original is variable and copies can be made which range from half the size to twice the size of the original, with an upper limit of 24 in. × 18 in. During exposure the original must be placed in a copy-holder to hold it flat, and must be well illuminated by lights. The readable negative can in many cases be used without the necessity of making a positive print. If, however, a positive is required, all that is necessary is to place the negative in the copy-holder and make a further copy from this. The *Photostat* is quite a large piece of apparatus and usually requires a small room to itself.

Some photocopying cameras available in the U.S.A. and in Europe are of a more advanced type. For example, the Haloid *Foto-Flo Recorder* is completely automatic including processing and drying, and also can make copies on both sides of the copying paper. Positive copies can be obtained with one exposure by the use of direct positive copying paper.

The photocopying camera is particularly suitable for copying library materials. Thick bound volumes can be held in the copy-holder and even though the pages curve in towards the centre of the volume, the camera lens usually has sufficient depth of focus to give a reasonable copy of the text at the inner margins.

Book-copiers

The essential part of a book-copier is the light-box which, as the name implies, is a rectangular box with some form of illumination inside. The electric bulbs or tubes which provide the illumination are so arranged that the light is spread evenly over the ground-glass screen which forms the top of the box. The screen itself helps to diffuse the light. To obtain the necessary contact between the original and the copying paper, most book-copiers are fitted with a device to provide mechanical pressure, which may be a lid with a pressure pad or a vacuum blanket. Book-copiers are usually fitted with a timing mechanism so that accurate exposures can be made. Some book-copiers are combined light-boxes and processing units but although such models usually cost less than when the two units are purchased separately, they are not necessarily easier to operate.

Many of the originals which are copied in a library are thick bound volumes, with tight sewing and narrow inner margins. It is essential therefore that a book-copier should be able to deal with this type of material. The ground-glass screen should be evenly illuminated and it should reach to the edge of the copier on at least one side. This will allow one page to be copied while the bound volume is held in the half-open position. In the case of a combined light-box and processor, the ground-glass screen should reach to an edge which is not above the feed-in to the processor. If this is not done the pages hanging down may be damaged by the chemicals.

The book-copier should have a method of applying mechanical pressure to the material to be copied. Whatever method is used any unevenness can be reduced by placing a plastic or thin card backing sheet immediately behind the page to be copied. The sheet should be inserted into the volume as far as it will go. Applying pressure by hand is not satisfactory as it can result in both spoilt copies and operator fatigue. The usual method is a lid which hinges down on to the original and then is locked into position with the pressure applied. If the lid has expanding hinges then it can be kept level and thus give even pressure on varying thicknesses of bound volumes. Where ordinary hinges are fitted the lid will only bring pressure to bear on the part of the volume nearest the hinge. If the volume is too thick for the lid to be fastened down the operator has to bring manual pressure to bear. This can bring about fatigue, but it can also have a more serious result. Although the manual pressure is applied at the edge of the lid, most of it is effective at the edge of the volume and may thus exert too great a pressure over a limited area of the glass screen causing it to shatter.

Pressure can also be applied with a vacuum blanket which is placed over the material to be copied and the air underneath the blanket removed with an electric pump. Once the vacuum has been obtained the normal air pressure holds the blanket and the material to be copied firmly against the ground-glass screen. When this method is used with flat documents, or unbound periodical parts exceptionally good contact is obtained. However when a bound volume is copied the edges of the text at the inner margins may be lost and the necessity to lay the entire volume flat on the glass screen may result in the binding being damaged. Pressure cannot be applied to a thick volume by this method because the blanket is held in a position where an air seal at the edges cannot be obtained so no vacuum can be formed.

Some copiers which have been designed principally for copying single papers have a device built in which allows the insertion of one leaf of a bound volume for copying leaving the rest of the volume outside the copier. A volume which has been copied by this method is never quite the same again. The pages involved tend to be eased out from the spine and to remain protruding beyond the other pages when the volume is closed. If the inner margins are narrow and too much pressure is applied the pages can easily be torn out.

The *Contoura* portable book copier is designed for manual pressure only. However it has a translucent plastic air-filled envelope which goes between the light source and the volume to be copied. This envelope will conform to any unevenness and thus gives even pressure over the whole page without the operator having to apply a heavy pressure.

The exposure setting device is normally calibrated in seconds so can be checked for accuracy and consistency. There are two types, one requires setting for every exposure, and the other need only be set when a change in exposure time is required. The latter is obviously the more convenient as the operator will for most exposures only need to press a switch.

Some method of varying light intensity is desirable in a book-copier. Many copying papers are available in different speeds and if a copier has a light intensity which is only suitable for one speed of paper it will be difficult to make prints on any other speed. For example if a copier has been designed to give good prints on a low speed paper at about fifteen seconds, a high speed paper will give a print in one or two seconds. On the face of it this would appear to be an advantage, but it means that the operator has little control over the exposure. If an under-exposed print is obtained in one second, the next setting of the timer at two seconds gives a 100% increase in exposure and this can result in an over-exposed print. Conversely a copier designed for high speed papers may take up to two and a half minutes to make a print on a low speed paper.

Copying papers

The sensitized paper used for copying purposes normally has a high contrast, which gives a dense black image on a good white background. Although this is ideal for copying textual matter, and will often produce copies that are better than the originals, it does mean that it is not suitable for copying illustrations. If for example a copy is made from a photograph, the darker grey tones in the original tend to become black in the copy, and the lighter tones tend to become white. This can be corrected to some extent if a half-tone screen is placed between the original and the copy paper when making the exposure. This breaks up the denseness of the blacks and gives more a acceptable copy; although, because it breaks the close contact between copying paper and original, it does cause slight loss of resolution in the copy.

The fact that copying papers give only blacks and whites means that coloured originals cannot be reproduced in colour by the normal process. This is, in many cases, not important, but if colour is used to denote particular information in the original then obviously the copy cannot be used to extract this information. Most colours appear on the copy as a shade of grey, but red appears as black. Because of this, black text on a red background or vice versa will appear only as a black area on a copy. Certain types of periodicals print a coloured background to parts of the text and it is difficult to produce an acceptable copy when the background is a shade of red.

The speed of a copying paper is of importance. As mentioned on a previous page the choice of the wrong speed of paper for a particular level of illumination in the book-copier can result in either too short or too long exposure times being required. Apart from this there is the question of handling the paper in ordinary room lighting. Placing the copy paper and the volume to be copied into position on the copier takes a certain amount of time. If a high speed paper is being used it could collect too much exposure to light during this time and would thus give a 'fogged' print when developed. If therefore the process is to be operated in ordinary room lighting a low speed paper is necessary, and even then the book-copier should be operated in a place that is well away from the windows and not directly under an artificial light source.

The latitude to exposure of a copying paper needs consideration.

If an acceptable copy can only be obtained by having an exposure exact within one or two seconds (this is known as highly critical exposure) then a fair amount of trial and error will be necessary and this will result in wasted copies. On the other hand if a paper has a wide latitude and acceptable copies can be produced anywhere within a range of say ten seconds, then there will be very few wasted copies.

In some processes several different weights of sensitized paper are available. Light weight paper is economical to use. Extra light weight paper can be used for making copies which are going to be sent by post or air-mail. Care has to be taken when developing these papers to ensure that they do not wrap round one of the rollers in the processing unit. There are also some different paper surfaces available. For example if maps or plans are being copied by the negative/positive process, the positive print can be made on to a rough surface paper. This surface accepts colour washes, so the map or plan can be coloured by hand if so desired. There is a glossy surfaced dyeline paper available which is particularly suitable for half-tone prints.

One further point in the handling of copying papers needs to be mentioned. There is a limit to the length of time that sensitized material can be stored without deterioration even though it is still unopened in its original wrappings. The safe period, which is known as the shelf-life, varies according to the material. Whatever type of copying paper is being used, care should be taken to ascertain its shelf life, and a form of stock control instituted to ensure that packets of paper do not lie on the shelves unused for a longer period. The normal shelf life of a copying paper can be reduced by incorrect storage. The store should be cool and dry and paper should never be kept anywhere near heating pipes.

The permanency of a photocopy is affected by the method of development used. If a copy is going to be in regular use for several years, a process must be used that produces permanent prints. If, however, a copy is wanted only for a short time, or, alternatively, is going to spend most of its life in a file with only occasional exposure to light, then the permanence of the print is not so important. Silver-halide papers if properly developed, fixed and washed will last many years without fading or discoloration. On the other hand, if a stabilization process without washing is used the prints will soon discolour.

COPYING WITH SILVER HALIDE PAPERS

1. *Negative/positive process*

With this process copies are usually made by the reflex method, but the direct method can be used with translucent single sided originals to produce readable negatives. The copying material is paper coated with a light-sensitive silver halide emulsion. After exposure the paper is placed in a developing solution which reduces the latent image to visible metallic silver. The unexposed undeveloped silver halide is then dissolved with hypo (i.e. in a fixing solution) and to ensure a permanent print a thorough washing is given to remove all soluble materials. Normal copying gives a copy opposite in tone to the original, so the first print will be a negative. The negative is then used by the direct method to make a positive print. The papers used in this process are sometimes referred to as document papers. They are available from several sources including Ilford, Kodak and E. N. Mason.

There are several light boxes available for this process, e.g. *Duostat*, *Contoura* etc. A combined light box and processing unit is not required as the development, etc., is done in open dishes. Near dark-room conditions are required for the processing stage, either very subdued lighting, or preferably orange safe-lights; and a supply of running water for washing the prints is a necessity. After washing, the prints can be pegged out to dry, but a small rotary drying machine will both speed the process and avoid the prints curling as they dry.

The negative/positive process has been speeded up by the introduction of a stabilizer which takes the place of the fixing and washing of the normal dish process. This means that the prints can now be developed by being fed rapidly through a processing unit where, by means of rollers, they receive a surface application of first a developing fluid and then a stabilizing fluid. Alternatively a single processing fluid may be used. This is referred to as a monobath. The prints emerge from the unit in a slightly damp state and rapidly dry out in normal room temperature. The negative can be used to make positives as soon as it is processed, there is no need to wait for it to dry. The one disadvantage of the stabilizer is some loss of permanency in the print. However this is not of major importance for most copies. Where a higher degree of permanence is required, the copies should be washed in running water for several minutes immediately after being processed. Examples of machines using this process are the Mason *Polyprint* and the Copycat *Dalcopier*. The *Polyprint* is only available as a rotary copier so will not make copies from bound volumes. The *Dalcopier* is available both as a rotary copier and as a book-copier combined with a processing unit.

2. *Direct positive process*

A specialised type of silver sensitized paper is available which gives a positive copy directly from an original, thus eliminating the intermediate negative when single-sided translucent originals are being copied. When copying opaque material a reversed or mirror positive is first made, using the reflex method, and then a normal positive printed from that.

A readable positive can be made in one step from an opaque original by means of what is sometimes called *indirect copying*. A sheet of direct-positive paper is placed between the original and another piece of ordinary white paper of reasonable translucency. The direct-positive paper has its sensitized surface away from the original and against the white paper. When the exposure is made the light passes through the two papers to the original where it is reflected from the white areas surrounding the text. The light then passes back through the direct-positive paper and some of it is reflected back on to the sensitized surface by the white paper. The resulting print has a dark grey text on a light grey background, which is only to be expected when the paper has been bombarded with light from all directions. It cannot be considered a good copy, but may be acceptable when a copy is needed quickly.

To get really good copies direct-positive paper should be exposed using a yellow light. Any light box can be used which has a high enough level of illumination, and a yellow filter can be placed over the ground-glass screen. Direct-positive papers can be handled in subdued indirect daylight or normal tungsten room lighting through all stages of printing and processing. There are some book copiers designed specially for use with these papers, and these usually have a processing unit available either as a separate item or combined with the light box. These processing units use the stabilizing technique already described.

Direct positive papers are marketed by several companies including Ilford (trade name *Inverpos*), Kodak (trade name *Auto-positive*) and E. N. Mason. Light weight papers are available and are particularly suitable for making dyeline masters.

Chemical Transfer Process

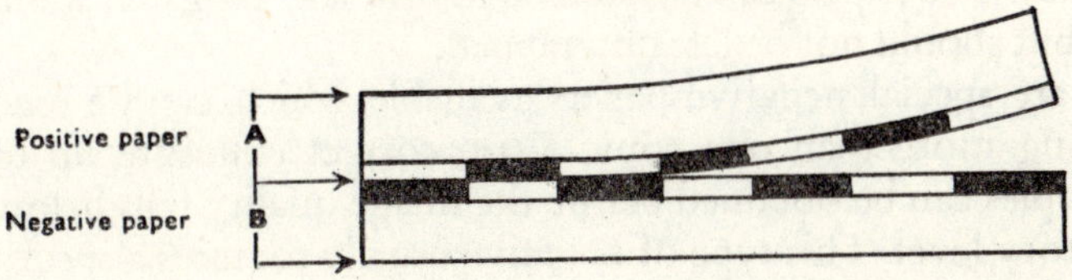

The Chemical Transfer negative paper is exposed to light in the same way as the Mechanical Transfer paper. After exposure it is passed through a developing machine together with a sheet of positive emulsion-coated paper. Having passed through the developer they emerge from the machine, having been pressed into close contact. During this contact the undeveloped silver salts in the surface coating of the negative paper migrate to the surface of the positive paper, where they are reduced and form a positive image.

Figure 6.

3. *Chemical transfer process*

This process, which is often referred to as the diffusion transfer process, requires the use of two emulsion-coated papers. The negative paper has a normal light sensitive coating, the positive paper has a coating containing developing nuclei. Exposure is always made by the reflex method. After exposure the negative sheet is placed (emulsion surface to emulsion surface) in contact with the positive sheet. They are then inserted together through adjacent slots in the processing unit, where they pass through a developing fluid and are then pressed firmly together by a pair of rollers. A negative image develops on the negative paper, and then the silver-halide in the unexposed part of the coating transfers under pressure to the positive paper where it is turned into black metallic silver by the development nuclei. Emerging from the processing unit they are left in contact for about half a minute and are then peeled apart. (See figure 6.) The transfer copy on the positive paper is then ready for use. The negative copy is normally discarded, although it can be used for making one further print if it is returned immediately to the processing unit together with a fresh sheet of positive paper. The second copy will not be a good one but in most cases would be acceptable. Several companies manufacture chemical transfer papers including Ilford (*Ilfacopy*), Gevaert (*Gevacopy*) and Agfa (*Copyrapid*). The papers can be handled in ordinary tungsten room lighting but should not be left unwrapped.

There are special negative papers available which can be used for making more than one copy. With correct exposure up to twelve copies can be obtained before the image quality falls below a satisfactory level. This type of negative usually requires a special positive paper to go with it. An example of this is the Gevaert *Multineg* and *Multipos* papers. An alternative and cheaper way of making several copies can be used when a dyeline process is available. Instead of transferring the positive print on to paper, a film is used instead. After transfer the film must be well washed and dried otherwise it will discolour. Once completed the film is used as a master to produce the required number of dyeline prints. The ordinary negative paper can be used to make this film, no special paper is required. An example of the film material is the Agfa *Copyrapid Film*.

The chemical transfer process can also be used to make offset plates. A special negative paper is used to transfer the print on to an aluminium offset plate. The plate can then be used in an offset printing machine to make several hundred copies. An example of this process is the Gevaert *Gevacopy Negative Offset* paper used in conjunction with a *Gevacopy Aluminium Offset* plate.

As the chemical transfer process is one of the most popular processes used for office copying there are many copiers available. However, as most of them have been designed to copy single sheets of paper, they are not suitable for copying bound volumes. When this process is for use in a library, then either a combined light-box and processing unit can be purchased, or alternatively they may be obtained as separate items. In the latter case the processing unit can be obtained as part of an office copier and thus used as required for single sheet copying, the separate light-box being used for making exposures when copying from books and other bound volumes. Examples of the combined book-copiers are the Ozalid *Combi*, E. N. Mason *Flatbed Photorapid*, Roneo *Flash Model 55*, Remington Rand *Transpeed All-purpose*, and Copycat *T.C. 1316*. Several separate light-boxes are available including the Block and Anderson *Unicop*, Anson *Copyfix Models 813* and *1316*, Copycat *T.C. 813*, Roneo *Flash Model 55X*, and the Duplomat *P.R. Printing Unit*. The separate processing units available include the Ozalid *Develop*, Duplomat *Developing Units*, Block and Anderson *Bandavelop*, and the Anson *Copyfix*. As far as the rotary office copiers are concerned their name is legion.

As many libraries use the *Contoura* light-box mention must be made of two processing units which can be used with the *Contoura* to make chemical transfer copies. The first is a small hand-operated unit (which can have a separate electric motor to drive it) called the *Constat*, the second is an electrically operated unit which is called the Anken *Attaché*, because it is made in the form of an attaché case and like the *Contoura* itself is easily carried in the hand.

4. *Mechanical Transfer Process*

In this process, which is sometimes referred to as the soft gelatine transfer process, the negative material consists of a paper coated with a gelatine emulsion containing a light-sensitive compound, a tanning developer and a dye-forming agent. After exposure by the reflex method the negative (usually known as the matrix) is placed in an alkaline activating solution where the dye is formed. At the same time the part of the matrix affected by the light, i.e. the part corresponding to the white background area of the original, is hardened, leaving the parts corresponding to the text in the original still soft. A sheet of paper is then placed in contact with the matrix and the two sheets are then pressed firmly together. During this contact the dye is transferred from the unhardened parts of the matrix on to the paper thus forming a

Mechanical Transfer Process

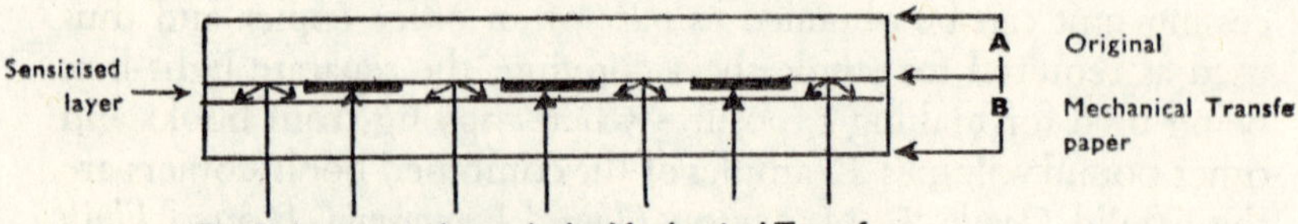

The light passes through the Mechanical Transfer paper and is reflected back by the non-text areas but absorbed by the text.

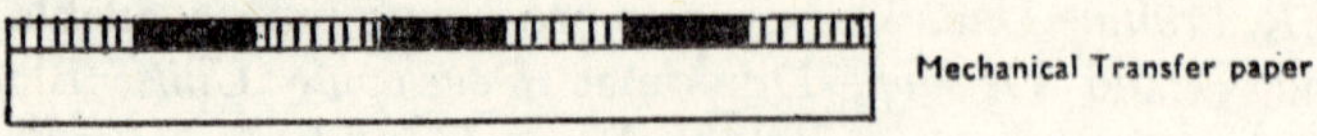

After exposure the paper is placed in an activating solution. This forms the dye and also hardens the exposed areas, leaving the unexposed areas still soft.

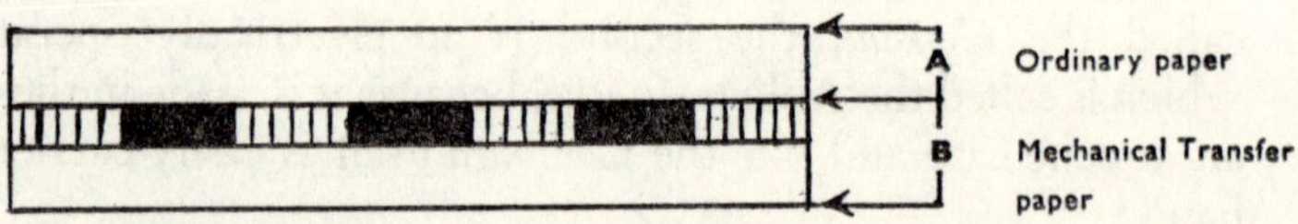

The Mechanical Transfer paper is then pressed into close contact with a sheet of ordinary paper

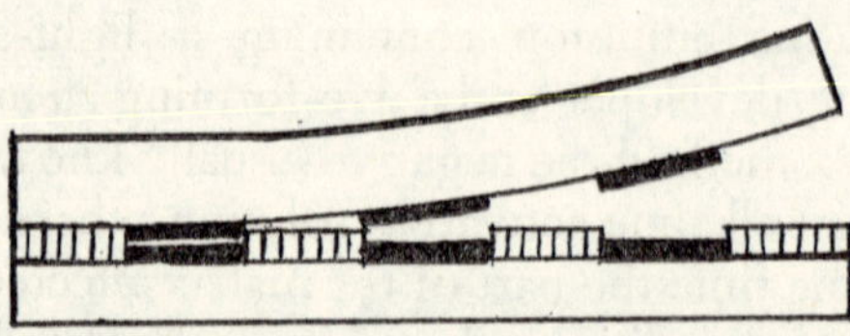

When the two papers are peeled apart some of the dye is transferred in the areas where the surface was soft. The result is a positive print.

Figure 7.

positive copy of the original The two are peeled apart and the transfer copy is ready for immediate use. (See figure 7.) The matrix can be used to make up to about six copies if the exposure was correct. It has to be re-inserted in the activating solution before being placed in contact with each copy. The transfer copies can be made on to ordinary typing paper, but better results are obtained by using a special paper which readily absorbs the dye.

The equipment for making copies by this process is manufactured by Kodak under the trade name *Verifax*. There are several types of copier each including the processing unit. The *Bantam* and *Signet* copiers are best for single sheets. The large *Verifax* copiers are suitable for books and single sheets. The *Verifax Witness* will copy documents up to 14 in × 18 in. in size. A *Verifax Offset Adapter* can be purchased separately and attached to the front of the copier. Using this device a copy can be transferred from a matrix on to a paper offset plate. The plate can then be used in an offset printing machine to make several hundred copies. Kodak also supply special equipment for making offset-plates under the trade name *Fktalith*. The matrix is of a higher quality and is much more light sensitive than the *Verifax* matrix. Fine details can be reproduced, solids are free from mottle and backgrounds completely clear. Used in conjunction with a Kodak Magenta Contact Screen this process can be used to produce half-tone prints from continuous tone originals such as photographs and sketches. Dyeline masters can be made by transferring from the matrix on to *Verifax Translucent Copy-paper*. A *Verifax* portable book copying unit is available. It has a carrying case and is fitted with an exposure compensator to allow for voltage variation at any copying location.

DYELINE (OR DIAZO) PROCESS

In this process the materials used are coated with a diazo compound which on development turns to a dye. Exposure to ultra-violet light destroys the power of the compound to form a dye. Thus when the exposure is made through a translucent original on to the dyeline paper, the dye forming properties are destroyed where the light passes through the original and are not affected where the printed areas of the original shield them. (See figure 8.)

Direct copying with Direct Positive paper or with Diazo paper

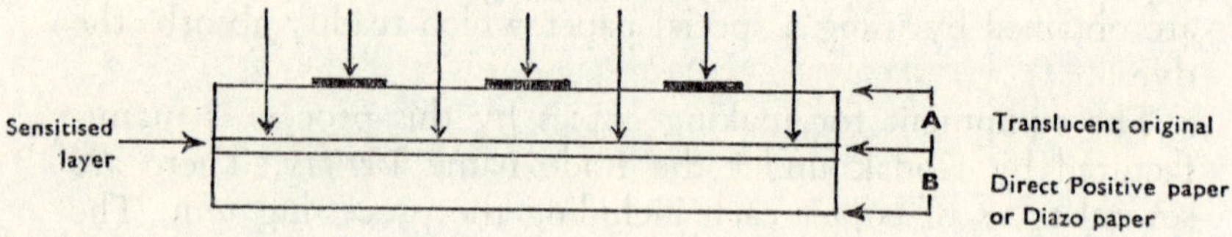

The light-rays pass through the non-text areas of the original but are absorbed by the text. When the Direct Positive paper or Diazo paper is developed the areas which have been exposed to light remain white and the areas which have been protected by the text turn black. The result is a positive print.

Figure 8.

After exposure development is in one of three ways depending on the type of dyeline paper used:-

(i) a semi-dry process in which the paper is passed through rollers moistened with an alkaline coupling solution.
(ii) a dry process in which the copy is developed by ammonia fumes.
(iii) a dry process in which the copy is developed by heat alone.

The dyeline process is quick and clean, and is ideal for the rapid copying of single-sided translucent material. It will not make copies from opaque or double-sided originals without the use of ancillary equipment to produce a translucent master. Once the master has been made any number of dyeline copies can be made from it. In the Ilford *Azoflex* process the master can be made on a *reflex foil* which consists of a transparent base coated with a diazo compound and covered with a fine carbon mesh. When the exposure is made the foil is placed in contact with the original with the mesh-covered side towards the light source. The mesh preserves the diazo compound from a large proportion of the direct light, but permits enough light to pass to strike the original and reflect back into the foil where it destroys the dye forming properties in the areas corresponding to the white background of the original. (See figure 9.) The mesh is then removed using a

Reflex copying using Azoflex Reflex foil

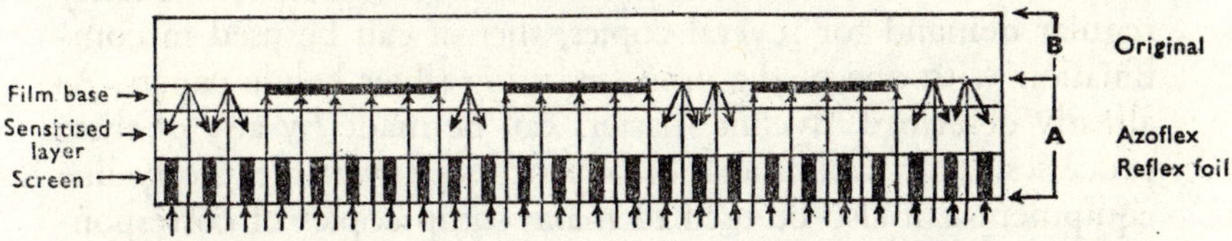

The light-rays pass through the meshes of the screen and are reflected back by the non-text areas of the original but absorbed by the text. The Diazo dye in the sensitised layer is destroyed where the light passes through it. Where it is protected by the screen the dye is preserved and is only then destroyed where the light is reflected from the original. Before development the screen must be removed. After development the foil is used as a master to print copies on Diazo paper.

Figure 9.

stripping device, and the foil is developed in a separate processing unit. The foil is then used on an *Azoflex* machine to make dyeline copies as required.

Masters can also be made by using *Azoflex transfer foil.* Like the *reflex foil* the *transfer foil* is coated with a diazo compound but beneath this is a layer of carbon black which acts as a screen and also provides the materials for the final image. Upon exposure, by the reflex method, hardening of the diazo compound occurs in the areas where the light has been reflected from the white background of the original. The foil and a suitable paper are passed together through an *Azoflex* transfer machine where they are immersed in ordinary water and then pressed firmly together. When they are peeled apart the carbon image from the unhardened areas, i.e. corresponding to the text of the original, is transferred from the foil to the paper. In its method this process is not dissimilar to the mechanical transfer process. If the copy is transferred to a sheet of translucent paper, this can then be used as a master for printing dyeline copies. If a considerable number of copies are required the transfer can be made on to an offset plate which can then be used to print the copies in an offset printing machine. There are also available pre-sensitized dyeline offset plates which are exposed by the direct method and after the use of one liquid developer can be used immediately in a printing machine (e.g. Ozalid *Pacemaster*).

The dyeline process is not an economic proposition for making single copies from bound volumes. However, if there is a fairly regular demand for several copies, then it can be used in combination with one of the processes using silver halide papers. As already described, dyeline masters can be made by any of these processes. If such a combination of processes is available, the dyeline equipment can also be used to make cheap copies of correspondence, invoices, etc. or even as a method of publication of, for example, reading lists, abstracts, index cards, where only a few copies are required.

Dyeline papers usually give a black or dark brown image, the latter requiring less exposure than the former, which allows quicker printing times, but having the disadvantage of a more critical exposure time which can result in a higher number of unacceptable copies. Other colours can be obtained but when the semi-dry process is being used the paper and the coupling solution must be matched to give the required colour. This means removing one solution from the machine and replacing it with another whenever a colour change is required. The Charles Bruning Company of the U.S.A. have developed a method for making multi-coloured prints by the dyeline process. The exposure is made in the copying machine in the normal manner, then small quantities of coupling solutions, one for each colour, are applied by hand, using cotton pads, to the appropriate parts of the undeveloped image. This is obviously not a process for multiple copying, but is very suitable for one or two copies of coloured graphs, diagrams, etc.

Because dyeline papers have a high contrast, photographs and other originals with a range of tones can only be successfully copied by using a half-tone screen to make a film master and then using the master to make dyeline prints. A film master can be made via the chemical transfer process, using a Kodak 120 line contact screen (magenta). This will give acceptable prints on ordinary matt-surfaced dyeline paper, but the use of glossy surfaced paper or J. Halden and Co. *Nuvatone* paper will give extremely good prints.

Dyeline copying machines are sold by several companies including Ilford, Copycat, Block and Anderson, Shannon, and Ozalid. Most of them are rotary machines so will only copy single sheets. This does not matter if the masters are being prepared by another process. If, however, dyeline masters of the *Azoflex* reflex foil type are to be used, then a light box with a strong ultra-violet light source is needed. In which case it could be advantageous to obtain a light box in which the intensity of the light can be varied so as to be suitable for both dyeline and silver halide papers, e.g. the Ozalid *OR510* plate-making, diazo and reflex unit.

Copying machines using the semi-dry process need to be sited near a supply of water for mixing the developing solution and for cleaning the developing trays. Those using ammonia fumes as the developing agent need to be fitted with ducts to remove the excess fumes, and great care has to be taken to ensure that this exhaust system is functioning perfectly otherwise the photocopying staff are apt to get tearful. The newer technique of developing by heat alone avoids the disadvantages of the other two methods. Heat developed papers are made by Ozalid (*Ozatherm*) and Copycat (*Diazotherm*).

PSEUDO-PHOTOGRAPHIC COPYING PROCESSES

1. *Thermographic process*

As its name implies this process uses heat instead of light to make copies. The sensitized material consists of a thin translucent paper used as a base for a plastic coating which contains two compounds that react under the influence of heat to form a dark-coloured compound. The document to be copied is placed in contact with a sheet of the copy paper. Exposure to an infra-red source is made by the reflex method. The infra-red light passes through the copy paper and strikes the original. In the white background areas it is largely reflected back still as infra-red light. However where it strikes the text it is absorbed by the printing ink and converted to heat. This radiates on to the copying paper and causes the sensitized layer to become a dark colour in the areas corresponding to the text on the original. (See figure 10.) Any document can be copied provided the printed or written matter has a carbon content. Copies can be produced from normal letterpress, pencil, typewriter, carbon copies, offset litho and stencil originals.

Thermographic process

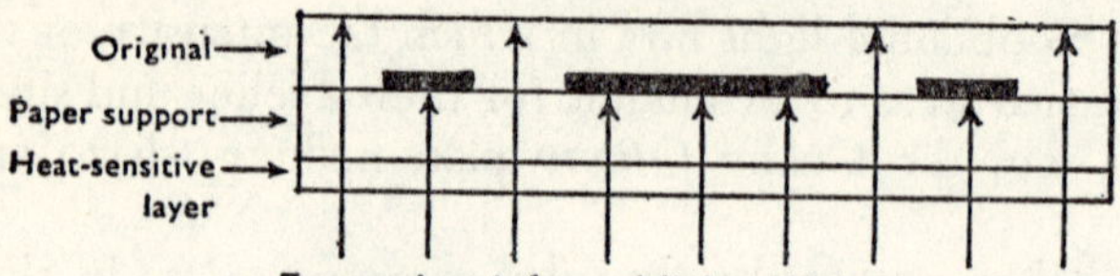

Exposed to infra-red light which heats the text of the original.

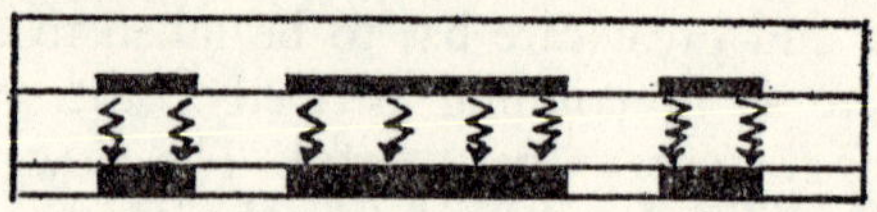

The heat reflects on to the sensitive layer which turns dark blue.

Figure 10.

Copies cannot be produced from originals prepared by spirit duplication, dyeline, several writing inks, coloured printing inks nor from thermographic copies. Pins, staples and paper clips have to be removed from an original before a copy is made.

Equipment for operating this process is manufactured by the 3M Company under the trade name *Thermofax*. The *Thermofax Secretary* is a rotary copier which will only copy single sheets. Books and other thick documents can be copied on the *Thermofax Premier*. In this latter the original and copying paper are placed on a glass screen in the usual manner. The exposure is provided by a tubular light which travels across the machine, the exposure time being determined by the rate of movement of the light. With continuous copying the light source in *Thermofax* machines gets gradually hotter and this naturally affects the exposure time which gets steadily shorter. This can result in wasted copies. However, the latest model of the rotary copier, the *Thermofax Secretary 44*, is fitted with a temperature gauge which is part of the exposure setting control and this allows the operator to make the necessary compensatory adjustments as required.

A different approach to the thermographic process is used in the Eichner *Drycopy* machine. The original, a sheet of ordinary office paper, and a sheet of *Drycopy* transfer paper are run through the machine together. The heat generated by the text of the original melts the coating on the *Drycopy* paper and an exact copy is transferred to the ordinary office paper. If the text on the original is fairly widely spaced, the transfer paper can be used to make another copy by using up the sections between the lines of the first copy.

Thermographic papers do not lose their sensitivity after exposure so should be kept away from heat at all times. If certain copies are expected to receive a heavy handling, they can be strengthened with backing sheets. The copy and a backing sheet are passed through the copying machine together and emerge firmly bonded one to the other. The thermographic process is very versatile and apart from straightforward copying the machines can be used to make offset paper plates, to laminate documents in a plastic film, to make black or coloured transparencies for use in a projector, to produce short runs of duplicated copies, to produce spirit duplication masters, and to reproduce lists of addresses on to gummed and perforated copy paper.

2. *Electrostatic process*

This process is based upon the fact that certain substances will retain an electrostatic charge whilst in total darkness but will allow the charge to leak away when exposed to light. The *Xerox* copiers (manufactured by the Haloid Company of Rochester, N.Y. and their British agents Rank-Xerox Ltd.) use a selenium coated plate or drum. The electrostatic charge is applied to this photoconductive coating and it is then exposed to light which has been reflected from the original and then resolved through a lens.

The charge leaks away in the areas affected by light from the background of the original, but remains in areas representing the text. Powdered resin particles are then strewn across the plate or drum and adhere where the charge is still in existence, thus forming a reversed image. A sheet of paper is given an electrostatic charge of opposite polarity and is then placed in contact with the plate or drum. The resin particles are attracted by the charge

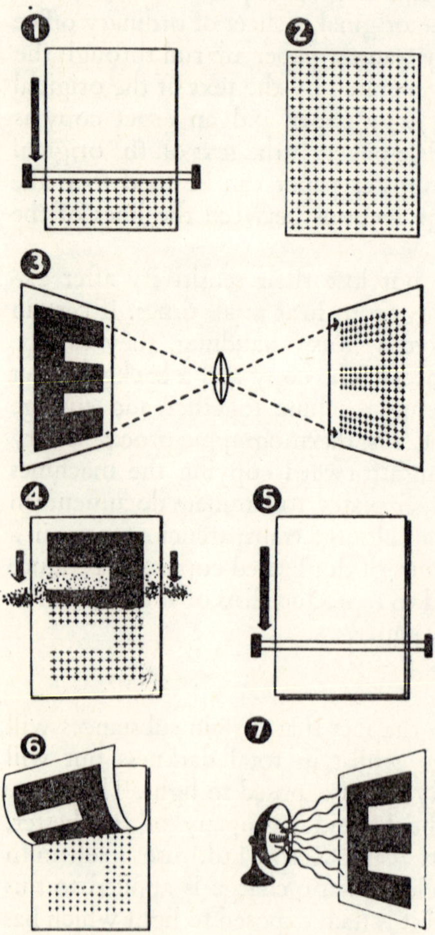

HOW XEROGRAPHY WORKS

1. Surface of specially coated plate is being electrically charged as it passes under wires. 2. Shows coating of plate charged with positive electricity. 3. Copy (E) is projected through lens in camera. Plus marks show projected image with positive charges. Positive charges disappear in areas exposed to light as shown by white space. 4. A negatively charged powder adheres to positively charged image. 5. After powder treatment (Fig. 4) a sheet of paper is placed over plate and receives positive charge. 6. Positively charged paper attracts powder from plate forming direct positive image. 7. Print is heated for a few seconds to fuse powder and form permanent print.

Figure 11.

and transfer to the paper. The paper is then submitted to heat which melts the powder image so that it fuses with the surface of the paper and becomes a permanent copy. (See figure 11.)

The *Electrofax* copiers (Radio Corporation of America) dispense with the plate or drum and use instead zinc oxide coated copying papers. The image is formed and fused directly on to the copying paper. This has the advantage of avoiding the use of the costly selenium coated plate or drum, but the disadvantage of using more expensive copying papers.

There is a considerable range of *Xerox* equipment in existence, but most of it is beyond the means of the average library even on a rental basis. However it may well be that some equipment is available within the organization or authority to which the library belongs and that a service may be obtained from it. The *Xerox* flat bed equipment *Model 1385*, can be used to make offset plates, dyeline masters and direct copies up to an image size of 13 in. × 8½ in. With the use of a *Xerox* variable ratio camera even larger documents can be copied, but will be reduced to not larger than the standard image size. To use this equipment the original is placed in the variable ratio camera or in the same-size camera. The selenium coated plate is charged in the processor and then placed in the camera and exposed to the original for a few seconds. The plate is then returned to the processor where the powder is cascaded several times over the surface of the plate. The reversed image is now visible on the plate and at this stage any unwanted matter can be carefully removed with a clean cloth. The offset plate, transparency or paper is placed in the processor in contact with the coated plate and the powder image is transferred. The plate or paper is then placed in the fuser where the heat is applied which makes the image permanent. A *Xerox Flo-Set* unit can be used in conjunction with this apparatus to make spirit duplication masters. If the powder image has been transferred on to a material which would be damaged by heat, e.g. acetate sheeting, then the *Xerox* vapour fuser can be used to melt the powder.

The *Xerox 1318* will make offset plates or dyeline masters up to an area of 13 in. × 18 in. This model is designed for copying engineering and architectural drawings and the plates and masters can be made either from the original drawings or by projection from microfilm. The *Xerox 1824* will provide enlarged copies of drawings from either negative or positive microfilm mounted in punched cards. There are several *Xerox Copyflo* models which will make copies at the rate of 20 feet per minute on a continuous roll of paper from original documents or from roll microfilm.

The *Xerox 914* office copier is a suitable model for library copying. (See plate 6.) It is available only on a rental basis and is economical where a copying service is making at least 1000 copies per month. The document to be copied is placed face downwards on a clear glass screen. The exposure control is set for the number of copies required and the 'print' button is then pressed. The document is scanned from below by a moving lens, the illumination being provided by moving tubular lights. The image is developed on a rotating selenium-coated drum and transferred immediately to a sheet of paper which then passes through a heat fuser and is delivered into a tray on top of the machine. The whole process from pressing the button to the receipt of the finished copy is approximately 30 seconds, but during continuous working the machine delivers copies at the rate of one every 10 seconds. This copier provides same size copies up to 9 in. × 14 in. so will handle most library materials.

CHAPTER 3

Microcopying

Microcopying is not such a new technique as recent developments would lead one to suppose. Experiments in the production of minute images started soon after the introduction of the basic photographic processes. The first known microcopies were made in 1839 by John Benjamin Dancer, a Manchester optician. These copies were made on to daguerrotype plates, but the image structure was too coarse, so the experiments were discontinued. In 1852, Dancer resumed work using the new wet collodion process. This fulfilled his requirements and he produced minute images containing groups of portraits within a circle only $\frac{1}{16}$ inch diameter. He produced and sold images mounted on to microscope slides. An essay on the possible uses of microphotography was contributed to the Encyclopaedia Britannica of 1857 by Sir David Brewster. This article was many years ahead of its time.

In France, René Dagron had developed a microphotographic process before 1860. He soon realized that images mounted in a cheap viewing device were a better commercial proposition than those which need a microscope as a viewer. He is best known, though, for his work during the Franco-Prussian war of 1870. Using his process, microphotographs were made of dispatches, and these were attached to the legs of pigeons and flown in and out of besieged Paris. Some of these dispatches are still in existence and consist of sixteen pages reduced from an original area of $8\frac{1}{2}$ feet × $6\frac{1}{2}$ feet to 2 in. × $1\frac{1}{4}$ in. One pigeon could carry eighteen microcopies containing altogether a total of between 36,000 and 54,000 messages. On receipt the copies were mounted between glass and projected on to a wall. They were at first transcribed by hand, but towards the end of the siege a method of photographic enlargement was used.

Although curios continued to be made there was very little advance in microphotography until the First World War, when it was found to be a suitable medium for espionage purposes. This use was further developed in the Second World War when the Germans worked out a method for concealing messages in the full-stops on postcards. This was one of the uses suggested by Brewster in 1857. This system did not go long undiscovered and the British counter-espionage had a busy time removing German messages and substituting their own versions. Another use found for microphotography was the 'Airgraph' or 'V' mail, where letters written on special forms were reduced on to film and flown to their destination. Enlarged copies were then made and sent to the recipients.

Commercial microphotography can be said to date from 1928 when Eastman Kodak introduced apparatus for the copying of cheques in banks. The business uses of microfilm gradually developed and it was only natural that librarians became interested in this medium. Microfilm was at first the only process, but microfiche was developed about 1936, *Microprint* in 1939, and the microcard, although known as early as 1938, was brought into prominence by Dr. Fremont Rider in 1944, with the publication of his book '*The scholar and the future of the research library*'. Since the war a wider range of uses for microphotography has developed, particularly in the field of information retrieval.

The main library uses to which microcopying may be put are:-

1. Storage of material which is rarely used but which must not be discarded.
2. Making available material which is too expensive or too rare to obtain in normal form.
3. Copying for security purposes.
4. Publication of material for which the demand is too small to justify book form.
5. Inter-library loans.
6. Information retrieval.

Before a library embarks on a microcopying programme a study should be made of the problems involved. The purpose of the programme should be clearly defined, and the quantity of material involved ascertained. There is no point in purchasing and setting up an expensive camera to microfilm half-a-dozen documents per year. The sort of questions which need answering are:-

1. Should the work be sent to a service centre or should a microcopying service be set up in the library?
2. What type or types of microcopies are required?
3. What cameras, readers, and reader/printers are required?
4. Should processing equipment be included in the service or should the processing work be contracted out.
5. Where can space be found to house the service?
6. Who will be responsible for it and will extra staff be required to run it?

It may be gathered from the foregoing that a microcopying service is not ventured upon lightly.

Microcopies can be divided into two main categories—microtransparencies and micro-opaques. The former are usually on a film base and are read by transmitted light, i.e. the light passes through the film on its way to the eye of the reader. The latter are usually printed on paper and are read by reflected light. Microtransparencies are made in the form of microfilm or microfiche. Micro-opaques also fall into two main types, those made solely by photographic means, and those in which the final copy is printed.

Microfilm

Microfilm is made in four standard widths: 16 mm., 35 mm., 70 mm. and 105 mm. The two larger sizes are used for special purposes, e.g. engineering drawings, copying of displayed sets of papers, preparation of microfiche. The two smaller sizes are commonly used for document copying, and most copying for library purposes is done on to 35 mm. Films can be purchased with or without perforations along the edges, and obviously the unperforated film has more space available for the images. The number of pages that can be copied on to a reel of film depends upon the size of the original document and the reduction ratio. A large reduction ratio may economise on film but the quality of the negative may be such that it is impossible to make a legible enlarged print from it.

35 mm. microfilm is usually supplied by the manufacturers in 100 feet rolls. On this length 800 exposures can be made at 'full frame' size (24 mm. × 16 mm.) and 1600 at 'half-frame' size (18 mm. × 24 mm.) The exposures can cover one page of the document, or, in the case of a book, the two adjacent pages shewn when the book is open.

16 mm. microfilm is also supplied in 100 feet rolls. The number of documents which can be copied on it depends, particularly with automatic flow cameras, on the length of the document. The amount can be increased by using a Duplex camera which photographs both sides of the document simultaneously and records the two images side by side across the film. These methods, which can store over 10,000 documents on one film, are ideal for copying loose papers such as invoices and cheques, but have little use in library operations where most of the material to be copied is in book form.

For making the negative exposures the type of microfilm used is usually panchromatic with a very fine grain emulsion. This is suitable for copying books, newspapers, manuscripts, etc., and gives a proper balance, in black and white, of a coloured original. Positive copies can be made from different types of film. The normal photographic film is fine grained and high contrast, and requires the usual full processing treatment. For quick copies diazo film can be used. Equipment is available for making diazo film positives of single frames particularly for insertion into aperture cards. Another material which can be used is the *Kalvar* microfilm.

The image in the *Kalvar* microfilm process is produced by exposure to ultra-violet or near ultra-violet light and then developing by heat. The process is thus completely dry, and as *Kalvar* microfilm is sensitive only to this particular type of light it can be handled for reasonable periods of time in normal room lighting. A further advantage of this is that *Kalvar* microfilm does not have to be used and then developed completely. As many images as required can be printed on to it and then others added later. This is very convenient when microfilming records, files, etc. which are being continuously up-dated. The image in *Kalvar* microfilm is produced in a different way to photographic film. The emulsion consists of photosensitive compounds held in a plastic material. When the compounds are exposed to ultra-violet light they decompose and set up a pressure in the plastic. The material is then developed by heat which softens the plastic so that the decomposed compounds form minute bubbles in the plastic. When the film is examined in normal room lighting only a faint pinkish image is seen. However when it is placed in a microfilm reader with a single light source a good black and white positive image appears on the screen. This is because the light is scattered by the minute bubbles in the plastic and what appears on the screen is in fact an enlargement of the shadows cast by the bubbles.

British Standard 1153:1955 gives recommendations for the storage of microfilm. Microfilm will deteriorate if it is not stored in the right conditions of temperature and humidity. If the film gets damp there is the possibility of mould growth on the gelatine, and if film gets too dry it becomes brittle and will crack easily. Microfilm should therefore be stored in a relative humidity of below 60 per cent and violent heat fluctuations should be avoided. To maintain the correct humidity level some form of air conditioning is necessary. Some storage cabinets incorporate chemical drying and humidifying agents, but this is not recommended because of possible damage to the microfilm. Damage can also be caused by residual chemicals, or by injurious chemicals in the atmosphere. The former can be guarded against by correct processing, the latter requires air conditioning.

Storage should be considered from the angle of how long the microfilm will be in use. Microfilms can therefore be divided into those of commercial permanence, i.e. those required up to a maximum of twenty-five years; and those of archival permanence that will be required indefinitely. With the latter all the requirements of proper storage must be observed and the films should be examined every five years and copies made of any showing signs of deterioration. In addition to this the negative should not be used for viewing; copies should be made for this purpose.

Storage cabinets for microfilm rolls are made by several companies. Each microfilm is kept on a spool inside a cardboard box 4 in. × 4 in. × 1½ in. The boxes are stored on edge in the drawers of the cabinet. The serial number of the microfilm and its title can be written on the upper edge of the box to facilitate finding. It has been recommended that microfilm should be stored horizontally on edge to prevent friction between film surfaces, but the convenience of vertical filing has been the rule in the design of storage cabinets. Cabinets vary from those designed especially for microfilms and which have shallow drawers with divisions of the right width to separate the rows of microfilm (e.g. the Roneo Microfilm Storage Unit and the Kodak Microfilm Cabinet) to those which have been adapted from cabinets designed to take 6 in. × 4 in. index cards. Remington Rand make small racks which will fit inside ordinary vertical files. There is also available steel shelving of the right size along which the boxes of microfilm may be stored like so many miniature books.

Although microfilm is normally stored in rolls, there are other methods of storage. The film can be cut into strips and inserted into transparent jackets which are made of stiff plastic. These microfilm jackets can be stored in catalogue or index drawers and in this format microfilm becomes as flexible as microfiche or microcard. Single frames of both 16 mm. and 35 mm. microfilm can be inserted into aperture cards. This particular type of card is designed for use in a punched card information retrieval system. The subject content of the microfilm frame is coded in the form of holes punched into the card. When information on a particular subject is required, the punched cards are passed through a sorting machine and those cards carrying information on the required subject are separated from the others. The microfilm inserts can then be studied in a microfilm reader, or enlarged prints can be made for the enquirer to take away.

In the *Microrite* system the microfilm is laminated on to a transparent film sheet. A small machine called the *Microrite dispenser* does the job mechanically, and the resulting product is in effect a microfiche but with the advantage that further matter can be added to the card at any time, thus permitting up-dating. A similar system entitled *Microprint-tape*, developed by Kodak has a microcard as its end product. It is available only for 16 mm. microfilm. The film is sent to a Recordak processing laboratory and is returned with a positive contact copy printed on to a roll of paper. On the back of the paper is a pressure sensitive adhesive protected by a backing strip. The roll is fitted into a special dispenser, and as it is extruded from this machine the backing is stripped off and the positive copies are cut to any required length and then stuck on to a card. The card can be any suitable size, so using this method, material can be microfilmed and converted to 5 in. × 3 in. microcards for interfiling with the library's stock of purchased microcards.

Microfiche

This type of microtransparency is sometimes known as microsheet or microfilm sheet and consists of a sheet of film containing a series of microcopies. Microfiche can be made in one of three ways. It can be contact printed from strips of microfilm, it can be an actual frame cut from microfilm (usually 70 mm. or 105 mm. film), or it can be made directly with the use of a step-and-repeat camera. It can be either negative or positive.

The images on microfiche are arranged in rows, and details of the contents are usually given at the top of each sheet in a form large enough to be read by the naked eye. As with any film care must be taken to avoid scratches on the surface emulsion. Microfiche are therefore usually protected between two sheets of glass when inserted into a reading device; and are kept in envelopes when filed. The envelopes are usually made from paper and have to be free from any chemicals likely to cause the microfiche to deteriorate. The top of the envelope is cut away at the front so that the title of the microfiche can be easily read. Some protecting envelopes are made of transparent material so do not have to be removed for viewing and thus protect the microfiche even when in the reader.

Microfiche does not have such a high storage content as microfilm, but has the advantage of being more flexible in that new material can be added by insertion at the correct place in the file. It has also proved to be very suitable for inter-library loan purposes, because a flat sheet of film is easy to transmit by post, particularly air-mail. The Microfiche Foundation, based at Delft, The Netherlands, is an international non-profit making institute which propagates the use of the flat transparent microfilm-sheet in science, technology, archives, etc. It will give advice on the use of microfiche and related equipment and also endeavours to keep a record of material published in microfiche form.

Microcards

The term *Microcard* is covered by an American patent which refers exclusively to 5 in. × 3 in. cards with images arranged in a specific manner. However, the term is used generally to describe any photographically produced card carrying a number of microcopies. The correct general term should be micro-opaque card.

The microcard resembles the microfiche in that the images are arranged in rows and bibliographic details are given at the top of the card. As microcards are positive prints from negative film they can be made by printing from 16 mm or 35 mm. film strips set in a holding device, or from sheet film made with a step-and-repeat camera.

Micro-opaques are made in various sizes. The *Microcard* is 5 in. × 3 in. and can carry up to almost 80 microcopies of original pages. The *Microlex* card is 8½ in. × 6½ in. and carries 200 microcopies. Photographically produced cards tend to curl and this has been overcome by laminating them back to back in pairs. Double sided cards of this type can be stored in approximately two thirds the space required to store them as separate single-sided cards. Standardization of card sizes is desirable for convenience in storage, and also that the design of readers may be simplified. The *Microprint tape*, referred to on page 59, allows the building up of one's own set of microcards on any size of card suitable for interfiling with existing sets. Similar services are available in the U.S.A. under the trade names *Microstrip* and *Microtape*.

Microcards are stored in standard catalogue drawers but *Microlex* cards need special cabinets. It was originally envisaged that the *Microcard* would be filed in place of the catalogue card thus allowing the reader to find the book at the same time as he found the catalogue entry. This has, in most cases, proved impracticable and *Microcards* are usually filed separately in their own cabinets. However there have been interesting developments of the original idea, one example of this being the publication of reports issued by the Commissariat a l'Energie Atomique as microcards but with English and French summaries in type legible to the naked eye.

Microprint

In this process the original is copied on to microfilm which is then used to make plates for offset-litho printing. The resulting product is a printed sheet 9 in. by 6 in. with 100 pages in micro-form, arranged in ten rows of ten pages each, on each side of the sheet. *Microprint* is economical to produce when fifty or more copies are required. Long runs keep the cost down and reduce it to approximately half the cost of microfilm. *Microprint* can be stored in vertical files or can be kept in pamphlet cases on the shelves, thus taking the place of the original work but occupying only 3 per cent of the space. As *Microprint* is in the form of ink on paper there is no question of fading or other deterioration as there is in purely photographic processes. The Readex Microprint Corporation have made special readers for *Microprint*, but they can also be used for other micro-opaques.

Cameras

The type of camera most suitable for a particular library will be the one that is best fitted for the programme which the library intends to carry out over the next ten years or so. There is no point in buying an expensive camera when comparatively few items will be copied. In fact there may be no point in buying a camera at all, for the work could probably be done more cheaply by a commercial firm on a contract basis. However if a large number of microcopies are to be made then the work can cost less and be kept under close control by having one's own camera.

The three basic actions involved in making a microcopy of a document are (a) placing the document into position in front of the camera lens, (b) illuminating the document, and (c) making a correct photographic exposure with the camera. Cameras vary from the simple where the three operations are performed manually, to the highly complex where nearly everything is automatic. The purchase price of a camera will often affect selection, but it can be a false economy to buy a cheaper camera which is slow to operate, because the labour costs in operating it may well total more than the purchase price of a fast camera.

To evaluate a camera the following questions should be asked:-

(a) Is the camera easy to operate, or does it require considerable technical skill? The simpler the operation the easier it is to train any member of the staff to use the camera.
(b) Are the settings for focus and exposure easily and/or automatically obtained? If not, then time and film will be wasted.
(c) Has the camera a good lens? If not the rest of the apparatus is of little use. The lens should have a minimum resolution of 110 lines per millimetre, should have all glass to air surfaces coated to reduce unwanted reflections, and should produce an image that is sharply focused over its whole area.
(d) Can short runs be made and the exposed film removed easily?
(e) Can more than one size of microfilm be used?
(f) Can microfilm without sprocket holes be used?
(g) Is there a warning device to indicate when the end of the film is reached?
(h) Are the light sources an integral part of the apparatus?
(i) Are there any auxiliary devices available, e.g. book-holders, card-holders?

The points are of varying importance, but the key items are a good lens and simplicity of operation.

Cameras for making microcopies are basically of three types, planetary or flat-bed, rotary, and step and repeat. The first named is suitable for microfilm copying of multi-paged documents, and can also be used to produce 16 mm. microfilm copies for conversion to microfiche form. The rotary camera is used to microcopy single sheets and is therefore not suitable for normal library microfilm work, but can be used for copying catalogue and index cards. The step-and-repeat camera, as its name implies, takes a series of photographs in a series of rows on a microfilm sheet. In other words it is a camera for making microfiche.

Of the many cameras available one of each type has been selected as an example for description in the following paragraphs. Their inclusion must not be taken as a specific recommendation although each is a good one of its type. No prices have been given because such information soon becomes out of date.

The *Recordak Micro-file* Camera, model MRD-2 (see plate 7) is a planetary camera which will photograph documents up to 26 in. × 36¾ in. in size. The reduction range is from 5:1 to 21:1 and there is a scale which enables the required reduction ratio to be selected. This is quite important because it is necessary to make the microcopies at a reduction ratio which is compatible with the reading equipment which will ultimately be used to consult the microcopies. All the controls are integrated into the base of the document table.

The camera takes the standard 100 foot rolls of 35 mm. microfilm but can be converted to take 16 mm. microfilm. The shutter has a fixed speed, and exposure control is obtained by varying the light intensity. The correct lighting is determined by a built-in electric cell exposure meter. The camera head will rotate in order to photograph documents horizontally or vertically along the film. The film advance for each exposure can be regulated in order to use only the exact amount of film required.

A light, projected through the camera lens can be switched on to illuminate the document table and define the exact area that the lens is covering. The area can be made larger by raising the camera and vice versa. A split-field range finder makes it easy for the operator to adjust the copy level through a 5 in. range for critical depth of focus. For actual exposure the document is illuminated by four 150 watt reflector flood lamps. Exposure is effected by a manual switch which operates both the shutter and the film advance mechanism.

The *Recordak Portable Microfilmer*, model RP-1 (see plate 8) is an easy to handle, rotary camera, weighing 24 lb., which can be taken in its own carrying case to any place where documents are to be microfilmed. This saves time and loss in moving material from its normal location. It takes single sheet documents up to

12 in. wide, and can copy as many as 2,500 letters on to a 100 foot roll of 16 mm. microfilm. If required it can expose simultaneously two rolls of film. It has a standard reduction ratio of 20:1.

The film unit can be easily interchanged if 'one subject' microfilming is required. There is a film footage indicator and a document counter. A special film spacing control allows separation between specific jobs; and there is a voltage control available as an accessory, which allows the operator to adjust the illumination in order to make good microcopies of coloured documents. A warning alarm sounds when the film is down to its last three feet.

The N.D.R. Goebel Camera is a step-and-repeat camera basically designed for microfiche although it can be supplemented with an interchangeable camera head for 35 mm. microfilm. A sheet of sensitized film is loaded into a slide which is then inserted into the camera. The material to be copied is placed on the base-board and the first exposure is made. The slide moves automatically to bring the next frame position on the film sheet to a point immediately behind the lens. This action is repeated until all the frames in the first line have been exposed. When the end of the line is reached a warning buzzer sounds, and the camera has to be returned to its starting position. As it is returned it moves to the next line ready for a further set of exposures.

The size of each micro-image, the number of frames to a line and the number of lines on the film sheet must be determined beforehand. A chart is supplied which enables the correct reduction ratio to be selected, after which the required reduction mark is placed between the lens and the film slide. There are two positioning racks for each reduction ratio. These have to be attached to the camera, where one sets the number of exposures per line and the other the number of lines per sheet. The camera can be moved up and down on its vertical supports and can give a range of reduction from 8:1 to 24:1.

The timing of the shutter and the lighting control are automatic and operated by a press button. Material for copying is placed on a moveable baseboard. Pressure on a foot pedal raises the board and flattens the material against a glass plate. After each exposure the baseboard is lowered for pages to be turned, or material exchanged.

The strips of exposures cover the film sheet except for a narrow strip at the top. When the sheet is fully exposed it is removed, still in its slide, and inserted into a titling camera. The title is typed or lettered on to a strip of paper and this is placed in a slot on the camera baseboard. It is then photographed on to the previously unexposed strip, after which the film sheet is processed in the normal manner and becomes a negative microfiche.

Microform readers

The optical system of a microform reader varies according to the type of microform being read, i.e. micro-opaque or micro-transparency, and whether the enlarged image is projected internally or externally. In a reader designed for microtransparencies the light is transmitted through the microfilm or microfiche, and then through the lens; whereas in the reader designed for micro-opaques the light is projected on to the microcard from whence it is reflected through the lens. As a certain amount of light is lost in this process, a micro-opaque reader needs to have a stronger light source than a micro-transparency reader. If the reader is designed for internal projection the screen on to which the image is projected is an integral part of the reader and can be either an opaque screen or a translucent screen. The former is usually set in the base of the reader and protected from external light sources by the body of the reader. The latter is usually fitted on the front of the reader and may have side screens fitted to protect it from external light or may have a tinted surface to cut down both glare and reflected light. Readers designed for external projection are of two types. The first projects the light directly from the lens on to any vertical surface which can be used as a screen, and the second is so designed that the lens directs the light upwards to a mirror which then reflects the light downwards on to the horizontal surface on which the reader is standing. This latter design is the one most used for small portable readers which can be stood on a desk or table and the enlarged image projected on to a piece of white paper or a blotting pad. There is one other type of reader in which the micro-image is read directly by examination through a magnifying lens. Such a method is suitable for quick reference to a microform but not for continuous reading.

In the evaluation of a microform reader the following points should be considered:-

(a) Is the screen protected from stray light?
(b) Does the screen tilt to allow the most comfortable position for reading?
(c) Is there good definition of the enlarged image, with even lighting over the whole screen and no 'hotspots'?
(d) Do the enlarged images remain in focus when the microform is being moved on from one image to the next?
(e) Is there a range of lenses available for the reader, and are the enlargement sizes of the lenses compatible with the reduction ratios of the microforms.
(f) Is the operation simple? The controls should be easily accessible and it should be easy to load and unload the microforms.
(g) Is the maintenance easy? The load and feed mechanisms should be easily accessible for cleaning and should be so designed that no damage is caused to the microform. The light bulb should be easy to replace and there should be efficient control of the heat generated by the bulb.

There is a wide selection of microform readers available for purchase. Those which are described in the following paragraphs are good examples but no specific recommendation is intended.

The *Recordak Archival* Reader is specifically designed to take 16 mm. and 35 mm. microfilm. The light source, the optical system and the spindles for holding the microfilm reels are all placed on top of the reader, and the images are projected directly downwards on to an opaque screen, which is held in the base and protected from stray light on three sides by the body of the machine. The screen is 18 in. square and has a matt finish in order to minimise glare. The winding handle, for moving the microfilm past the light source, is fitted on the side of the reader. There are two lenses, a 24X magnification and a 12X magnification, and they are easily interchanged by the movement of a control on the reading head. The reading head itself can be rotated so that the image is always in the correct reading position. A scanning control knob permits movement of the image from side to side so that any part of it can be read conveniently. An accessory holder can be fitted beneath the lens in order to allow the reading of aperture cards.

The *Recordak Lodestar* Reader is an internal projection type of reader with a translucent tinted screen (see plate 9). It is designed for rapid reference to pre-indexed 16 mm. microfilm. The film is held in special magazines and has been visually coded in serial number order at the time it was made. The insertion of the magazine into a slot on the side of the reader automatically switches it on. The single control switch permits high, intermediate or slow scanning of the film. The index marks on the film are printed in between each micro-image and are read against the index scales on each side of the screen. There is no take-up spool, the film feeds into a special chamber where it festoons and is held until rewound into the magazine. Threading of the film through the reader and into another spool is completely eliminated and several magazines can be consulted one after the other at a rapid speed. The film can be passed through the reader at speeds up to 600 feet per minute, and a full re-wind can be completed in 10 seconds. There is a reader/printer model also available.

The *Dagmar Super* is a portable reader for 16 mm. and 35 mm. microfilm, for microfiche and for aperture cards. The reader folds down, and can be carried and/or stored in a compact case. The illumination source is inside the reader, and the light from it passes through an aperture in the outer case, through the microcopy, and then is projected through a lens on to an external mirror which is positioned on supports above the reader. The image is then reflected downwards and is focused on to a horizontal screen lying on the surface of the desk. The screen can be a piece of plain white paper or a blotting pad. Alternatively the reader can be tipped on to its back and the image projected on to a wall or on to a vertical screen. The mirror supports are telescopic and thus provide variable magnification.

The reader is fitted with removable reel arms for use with microfilm. Aperture cards and microfiche are held in a special glass holder, which is held in the reading position on the front of the reader by magnetism.

The *Microcard Corporation Micro III* Reader is a completely portable model which weighs only 6 lb. and when closed for carrying is approximately 13½ in. long, 9 in. wide and 3½ in. deep. It has a carrying handle at one end, which will also, if so desired, raise the front of the reader when it is in use. The screen which is 8 in. × 9½ in. in size, hinges upward for use and folds down inside the case afterwards. There is also room inside the case to carry a limited number of microcards. The lens gives an 18X enlargement. The micro-opaque to be read is held in position on the front of the machine by a metal clip, and is moved manually from page to page. There is also available an accessory clip-on film illuminator which enables microfiche and unitised microfilm to be read as well.

The *Ross Microreader* is a combined micro-opaque and micro-transparency reader and also has a photographic attachment for making enlarged prints. Basically, the reader is designed for use with microfiche and microcards. The microform to be read is sandwiched between two optical-glass plates, placed on the top of the reader and held in position by a spring clip. The microform is moved by hand to the required frame, and focusing is done by means of a sliding control lever. For reading microfilm a separate attachment with reels is fitted to the top of the reader. The optics are such that the illumination required is only one 12 watt bulb for microtransparencies and two 12 watt bulbs for micro-opaques. The transparencies are illuminated from above and the opaques from below, changeover from one source to the other being controlled by a switch.

The standard lens supplied with the equipment has a magnification of 15X which can be increased to 18X by use of a special screen extension unit; and alternative or additional lenses are available with magnifications of 10X, 27X and 37X. Approximately 22 lb. in weight the *Microreader* is reasonably portable and the manufacturers also make a carrying case with provision for accessories. A further optional item is the photographic attachment. This is fitted to the front of the reader in place of the screen. The image is properly focused by use of a small hand-held, ground-glass screen. A sheet of photographic paper is then placed in the attachment and held in position by a platen. The exposure is made by switching on the normal reader illumination. The photographic paper has then to be developed, fixed, etc. which requires darkroom conditions with the appropriate safelights.

Microform Reader/Printers

In the evaluation of reader/printers the same points must be considered as those for microform readers, listed on page 66. These criteria will take care of the reading aspects and the various controls. The printing requirements can be given very briefly. What is wanted is a quick automatic process that can be operated in normal room conditions and will provide a dry or semi-dry enlarged copy. Processes in use in reader/printers include the silver halide stabiliser process, the diazo process and the electrolytic process. The first two of these have already been described in Chapter 2. The last-named is a process developed by the 3M Company for use in their reader/printers so is best described here.

In the electrolytic process the sensitised material consists of a paper coated firstly with a thin layer of metal foil, and then with a thin layer of zinc oxide. The zinc oxide coating has a high electrical resistance when held in the dark, but when it is exposed to light this resistance is lowered. Consequently if it is exposed to light transmitted through a microfilm negative the resistance remains high in the non-text areas. After such an exposure the sensitised material is brought into brief contact with a sponge soaked with an electroplating solution, and at the same time a direct current is applied which passes between the solution and the metal foil layer. This causes a metallic deposition from the solution on to the zinc oxide, forming a visible image of the text. The surface of the sensitised material is only slightly moistened by the sponge, and it soon dries in normal room temperature.

The 3M Company *Filmac* 100 is one of a series of microform reader/printers which use the electrolytic process. It takes 16 mm. and 35 mm. microfilm, and projects a small but clear picture on to a 7 in. × 9 in. ground glass-screen which is placed on the face of the machine. There are five lenses available giving magnifications of 7X, 9X, 13X, 19X, and 26X. Two lenses can be mounted in the reader at any one time and the changeover from one to another is a simple operation. To obtain a print the required frame is projected on to the screen. The intensity control and the exposure time are then pre-set. The print button is pressed, and this sets into motion a series of operations. Firstly the internal mirror which is reflecting the frame on to the screen moves to a position in which the frame is projected on to the sensitised paper which is held in the top of the reader in the form of a roll. After the pre-set exposure time expires the mirror returns to its original position and the enlarged image re-appears on the screen. The exposed paper is moved forward and passing over the sponge holding the developing solution is ejected from a slot above the screen on the face of the reader. The enlarged print is torn off using the slot as a cutting edge. The print is slightly damp but soon dries.

The 3M Company also make a *Filmac* 100 microfiche reader/printer which is very similar to the microfilm model. It takes microfiche up to 8 in. × 5 in. in size.

The *Recordak Magnaprint* reader/printer (see plate 10) is designed to accept 16 mm. and 35 mm. microfilm in roll form and is readily adaptable to film magazine and to microfiche, unitised microfilm and aperture cards. There is a motorised drive for film in roll form or in magazines. The screen is 11 in. × 11 in. in size and is tinted green to reduce surface reflections. There are guide lines printed on the screen to enable the enlarged image to be easily orientated to the correct position for a print to be made. There are five lenses available with the following magnifications:- 11. 8X, 19X, 22.5X, 30X and 38X.

The enlarged prints are made by the silver halide process using a stabilising monobath solution. The sensitised paper is supplied in roll form and is available in three widths, 4 in., 5½ in., and 8½ in. To make a print the selected frame is first focused on to the screen, an intensity control is pre-set and the print button pressed. The printing cycle is fully automatic. The image is first projected on to the sensitised paper, the optics of the equipment being so designed that in order to obtain maximum print quality there is a direct projection without any intervening prisms or mirrors. After approximately 10 seconds the image returns to the screen. The exposed paper is cut from the roll and moves through the monobath where it is processed and then is extruded from the slot in the lower front of the machine. The print is damp but soon dries in normal room temperature.

Microform enlarger/printers

There are a few machines which are designed to produce enlarged prints from microform but which do not have the facility of a screen for continuous reading. The requirements of such machines are that they should be able to produce quickly good permanent prints by a dry or semi-dry process. The three examples chosen for description have little in common other than the facts that all are enlarger/printers and can be used for library purposes.

The Xerox *Copyflo* continuous enlarger printer is a highly sophisticated and expensive piece of machinery which will make enlarged prints from 16 mm. or 35 mm. microfilm on to a roll of paper, 11 in. in width. The reels of microfilm are fitted into the printer, the controls set, and the enlarged prints come rolling out at the rate of 1200 feet of paper per hour. Alternatively the enlarged prints can be made on to offset-master material and used as masters to make printed copies. Few libraries are likely to have sufficient work to justify owning or renting the *Copyflo*, and arrangements can be made to do the occasional job at a Xerox service centre. The obvious use that comes to mind is the preparation of a bound book of Xerox copies made from a microfilm of the original book.

The *Microcard Copier Type* 1 is specifically designed to produce an enlarged print from a frame on a microcard. The chemical transfer process is used and thus provides a positive print from the positive micro-image. The positive and negative papers are supplied in roll form and fitted in the machine. To obtain an enlarged print the microcard is fitted into the card holder and moved into the correct position which is checked through an enlarging viewing lens. The exposure timer is pre-set and the print button pressed. After about half a minute the cover is lifted and the sandwiched negative and positive sheets are taken out and peeled apart, the negative being discarded.

The Caps-Jeffree dyeline enlarger (see plate 11), as its name implies, makes dyeline enlargements from positive microfilm. The microfilm is placed in the reading head and the desired frame moved into position. The image is projected upwards on to a mirror which is supported above the machine by two tubular arms. The mirror reflects the image downwards on to a table which is an integral part of the machine. A sheet of dyeline paper is placed on the table, exposed to the projected image and then developed in a separate developing machine. The size of the final print can be varied from 10 in. × 8 in. to 30 in. × 20 in., the mirror being moved up or down, in a power assisted operation, to obtain the required size. The lens moves automatically in conjunction with the mirror so that the projected image is always in focus. The reading head will take roll microfilm and aperture cards.

CHAPTER 4

The Uses of Document Reproductions

This chapter is not an exhaustive survey of all the possible uses of document reproduction in libraries. It covers the principal ones, with outline descriptions of the ways in which the various processes can be used, and only gives technical details where necessary, in order to avoid as far as possible repetition of the material in the earlier chapters. New ideas are being developed all the time and the information given in this chapter should be used as a guide line to future possibilities.

The uses are divided into two main categories, service to readers and administrative purposes. The first named, being the larger category, has been separated into: direct service; catalogues, bulletins, etc.; and information retrieval. These sub-categories are not mutually exclusive but do provide a convenient breakdown.

Direct services to readers

The major use of document copying apparatus is the provision of photocopies of printed materials, usually from the library's own collections. Some libraries also supply microcopies, and as the personal ownership of microform readers becomes more general, this may well develop into the major service. The purpose of this service is to save the reader having to make lengthy transcripts of the material he requires for studying purposes. The provision of copies in any form is governed by the requirements of the Copyright Act, the details of which are given in Chapter 6.

The straightforward copying of textual matter rarely presents any problems, but there are some processes which present difficulties when copying half-tone or full-tone illustrations, particularly when such illustrations are on the same page as textual matter. Silver halide papers will provide good copies of illustrations if given the correct exposure, which is usually less than that required for text. If the illustration and text have to be copied together, the reduction of exposure can be obtained by cutting a piece of thin translucent yellow film to the correct size and placing it over the illustration during the exposure period. With the xerographic process, a white dot screen placed over the page to be copied during exposure, will provide in many cases an acceptable copy of an illustration together with legible textual matter. If dyeline copies of photographs are required, a translucent master should be made, either on positive film using a half-tone screen, or by the chemical transfer process using a line screen positive film. This master is then used to print the required number of copies, which can be on normal dyeline paper, or for better effect, on glossy surfaced dyeline paper.

Microforms can be used in several ways to raise the level of service. Firstly by building up the library's collections by the purchase of material available in microform. Such purchases may range from microcopies of theses, and other documents which are available only in microform, through to sets of periodicals which are not easily available as originals, or which have sufficient use to justify purchase but not to justify the allocation of a large amount of shelf space. Microcopies of rare or unique originals should be considered for purchase, especially perhaps the colour microfilm copies of illuminated manuscripts. The library can preserve too its own rare originals by providing microforms for consultation instead. In many cases the microform will be as useful as the original, and in some cases, e.g., microcopies of newspaper files, it will be more convenient and easier to handle than the original. The cost of making copies of one's own originals can be offset by making extra copies from the master negative as required, for sale to other libraries and to individuals. However, in making and selling such copies the requirements of the Copyright Act must be observed.

The use of teaching machines in programmed learning is rapidly developing at the moment, and in the same way that some libraries now provide taped language courses for use in the library, collections of programmed courses will, no doubt, be built up and batteries of teaching machines will occupy the desks in the study carrells.

Programmed learning is based on the principle of the learning cycle. Firstly the student is required to direct his attention to a limited amount of information at any one time. Secondly he is required to take action on that information, usually by providing an answer to a question. Thirdly he is given an immediate check to his answer and should proceed further only when he proves by giving the correct answer that he has assimilated the information. This cycle is repeated for each part of the instruction. Programmed learning can be presented in book form, on tape or record, or on microfilm. When microfilm is used each particular item of information is put on to one frame of the film and the film is read on a special type of microfilm reader. After reading a frame, the pupil responds to the question by pressing one or other of the buttons at the side of the screen. If he gives the correct response he proceeds to the next part of the instruction, otherwise he is presented with a frame which draws his attention to the information which he has misunderstood, and then requires him to press a button which will restore the original frame to the screen so that he can study it again. Information on the availability of programmed learning can be found in *Programmes in Print*, published annually by the Association for Programmed Learning.

Inter-library and postal loans

The advantages of supplying photocopies or microcopies in lieu of the original are:-

(a) the original stays on the shelves and thus remains available for consultation
(b) there is a saving in both postage and packaging costs
(c) the copy can be retained by the requesting library or reader.

The National Central Library application form has a box on it, which is marked with an X if a Xerox print is acceptable in place of the original, and an M if a Microfilm is likewise acceptable. The Chemical Society Library application form clearly specifies the economic aspects by including the phrase (which can be deleted if so required) 'If not available on loan, or if the cost of carriage would be greater than the cost of a photocopy, please make and supply a photocopy of:-'.

Microcopies are particularly suitable in place of the original when the request for loan has come from abroad. The compactness and low weight/information ratio of a microcopy means that it can be sent by air-mail and thus satisfy the demand as quickly as possible. The Library of the Technische Hogeschool of Delft in the Netherlands has pioneered this sort of service with microfiche. The only drawback from the user's point of view is that he will have to read the microcopy on his own library's reader or else ask to have enlarged prints made. This problem will however tend to dissappear, for with the increasing use of microforms it is a reasonable probability that the personal ownership of portable microform readers will become as widespread and commonplace as the present-day ownership of portable typewriters.

Catalogues and indexes

The production of multiple copies of catalogue and index cards can be effected by several processes. The process chosen should be one that gives a clear print and a near permanent image directly on to cards. Some processes are only economical if the entries are produced several at a time on to a sheet, which is then cut up into individual entries. The disadvantage of this method lies in the difficulty of accurate cutting of the cards from the sheet, and even minor differences can make the cards difficult to consult when filed in a catalogue drawer. The equipment for operating the process should therefore be capable of handling cards, particularly 5 in. × 3 in. cards, and also of producing the entry on the card in a consistent position. Some types of feeding mechanisms seem incapable of doing this and the position of the entry can vary from side to side of the card, and also can come out lopsided.

Whatever process is used, it is important to devise a form of standard catalogue or index entry in order to avoid as far as possible the need to type headings on to the cards after they have been printed.

The stencil duplicating process can be used to print cards in two ways. Firstly there is the stencil addressing system in which the card-framed stencils normally used to produce addressed envelopes, are used instead to produce catalogue entries. The stencils are cut on a typewriter, but usually need a special platen fitted to the typewriter to give them the necessary support while being cut. Secondly there is the ordinary stencil duplicator which can be adapted to take small stencils. An example of this is the Roneo 865 which has a special drum in which there are ink-feed perforations only in an area 5 in. × 3 in. The machine can be fitted with a powder spraying device for separation of the cards after printing in order to facilitate drying. This also allows the use of ordinary cards instead of the semi-absorbent cards which are somewhat thicker and take up more room in the catalogue drawers. Roneo also supply an A4 size stencil which has four 5 in. × 3 in. card positions pre-printed on it. After the entries have been typed for each card, the stencil is cut into four and each of the parts used in turn to print the required number of cards.

Catalogue cards can be produced by spirit duplication also, by using a spirit addressing system or by an ordinary spirit duplicator. There are a range of colours available including black, but none of them produce permanent prints. This may not matter if the cards are always kept in the catalogue drawers and thus protected from the light except when being consulted. If however they are likely to be openly displayed, and particularly near windows, then there is a chance that the image would be bleached away by the action of the light. The colour which is most resistant to this is purple.

The offset lithography process produces good permanent prints on to catalogue cards and the printing ink dries very quickly on the card, so that it is practically dry as soon as the card leaves the machine. The offset paper plate is the most economical to use and is satisfactory for this short run work. Cards can be typed four up on a single paper plate, or small plates suitable for one card can be obtained. Offset-lithography machines tend to be somewhat expensive to buy, and would not be economical used solely for card printing. However they are eminently suitable for the printing of library stationery (forms, tickets, etc.), and the overall use of such a machine may well justify the purchase.

Cards can be printed from embossed addressing plates made of either metal or plastic. The plate is prepared in an embossing machine and is then transferred to an addressing machine where the cards are printed by being pressed against the plate with a large carbon ribbon, similar to a typewriter ribbon, sandwiched between them. The plates can be corrected by having letters blanked out and then re-embossed. A major advantage of using this process is that the plates can be stored and used again as and when required. This makes it particularly suitable for a large system buying multiple copies of books over a period of time for location at several libraries each of which has its own catalogue.

Photocopying processes can be used to make catalogue cards, but care must be taken to see that the end product is acceptable. Silver halide and transfer papers are not particularly suitable for catalogues. Sensitised dyeline cards can be purchased and while they are very suitable for cards which are for temporary use, e.g. cards used for information dissemination, they are not really permanent enough for cataloguing purposes. Xerography gives good permanent images on cards and can do so by two methods. The prints can be made directly from typed or written originals on to sheets of card, which are then cut up to form the individual cards. Alternatively the originals can be microfilmed and the microfilm then used in the *Xerox Copyflo* to print on to a continuous roll, which is afterwards cut up into the individual cards. The union catalogue in the National Library, Ottawa, was compiled by this latter method. The existing catalogues of the principal libraries in Canada were microfilmed. The cards were made from the microfilm and then interfiled to form the union catalogue.

The use of electric typewriters which produce, and can be controlled by, punched paper tape is slowly developing in this field. As each catalogue entry is typed a paper tape is automatically punched. It can then be fed back through a reading head as many times as required to produce the extra cards; and can also be stored in order to make cards, or bibliographies, or catalogues in book form at a later date. The *Dura Mach* automatic typewriter has interchangeable typing heads so can be used to put special information, e.g. chemical structures, on to index cards. It will also produce its input record punched on to a card which can be indexed and filed. Although such a card takes up more room in filing than a strip of paper tape, it is more convenient to retrieve and to handle.

Printed catalogues, Union lists and Bibliographies

The setting up in type, or typing on to offset plates, of each edition of a printed list of any kind is time consuming and expensive, and high additional costs are incurred by the very necessary proof-reading involved. Consequently it is not surprising that a number of techniques have been developed for preparing several editions of lists with only one typing; or, where the typing has already been done for another purpose, using this original typing to produce a list.

If it is known that a list is to be printed regularly it can be prepared by typing each item on to a strip of paper which is stored in a special holder. An example of this is the Kalamazoo *Copystrip*. The strips are available in widths to take one, two, or three lines of typing and are made up into sheets so that they can be held in a typewriter. After typing the strips are cut off the sheet using a special guillotine, and are slotted into pages made of a white plastic material. For convenience the pages are held in a loose-leaf binder. New additions to the list are slotted in at the correct place. When a page is filled a new page is inserted in the binder alongside the full page and half the contents transferred to it. When a new edition of the list is to be printed, the strips are made up into full pages, these are detached from the binder and put on to offset-litho plates by a photographic process or by Xerography. The plates are then used to print the list. No proof-reading is required because if the original strips were carefully checked they must be correct.

The use of embossed addressing plates as printing masters for cards, also allows them to be used to prepare lists. The required plates are fed through the addressing machine and are printed as a list on a sheet of paper. Care has to be taken that the paper is kept straight during this part of the process. Alternatively, to avoid this problem, the plates can be printed on to thin cards which are then stuck on to sheets of paper so as to form a list. In both cases the sheet of paper is then used to make an offset-litho plate which in turn is used to print the list. Similarly if a tape operated typewriter has been used to prepare the cards, the tapes can be fed through the typewriter to produce instead a list on paper for use in the making of the offset-litho plate.

Another method of preparing lists is the use of a listing camera. The entries are typed on a strip at the top of a suitable white card, usually machine-punched card size. The cards are sorted into order and placed in the listing camera where they are fed across the lens at a fairly high speed and the strip bearing the typing on each card is photographed on to sensitised film (or paper). After development the film can be used to make offset-litho plates or to make a positive film for use as a dyeline master. This process is particularly useful in the production of holdings lists of periodicals. The title, and other details, of each periodical is typed on to one card, and a serial number representing the title's place in the list is punched into the body of the card. The holding of each library, together with the code letters or numbers of the library, is typed on to a separate card, and the serial number of the title and the code for the library are punched into the body of the card. Normally all the cards will be filed in correct sequence, each title in correct alphabetical order followed by the cards representing the various libraries' holdings. New entries are filed as and when made. Each time a printed version is required the cards are fed through the listing camera and the offset-litho plates are made and used for printing the list. If one of the participating libraries require a list of its own holdings, the cards are passed through a punched card sorter and the appropriate cards extracted and used via the listing camera to prepare the required list. The sorter is then used to merge the extracted cards back into the main sequence. Also if by any mishap cards get misfiled the set of cards can always be put into correct order by using the sorter. The *Combined list of periodicals holdings in the libraries of Imperial Chemical Industries Limited* is prepared this way annually, and some of the Divisional libraries also have annual print-outs of their own holdings.

In cases where the information to be listed already exists in another form there are techniques available which allow the original typing or printing to be used in the preparation of a list. The two obvious library uses of such techniques are where a

PLATE 1

Rotary stencil duplicator (Roneo 865).

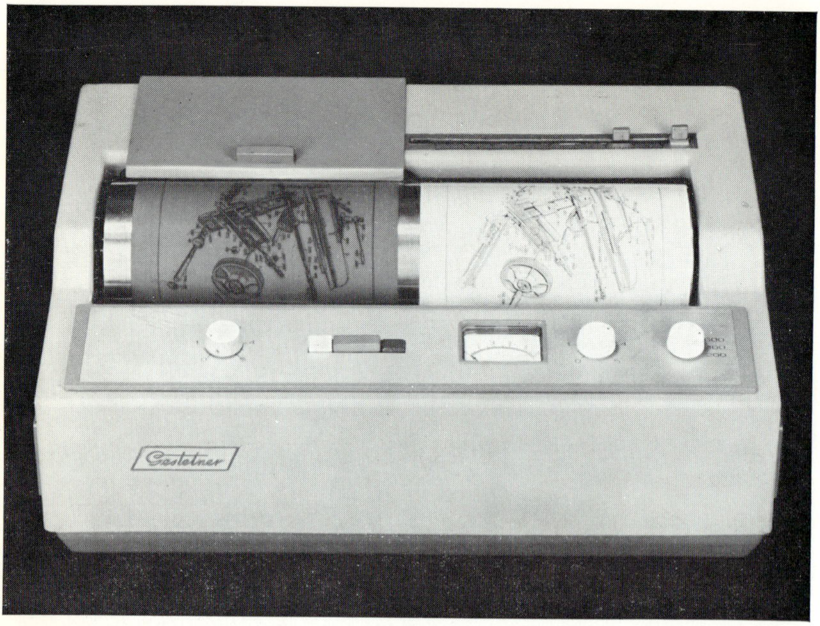

PLATE 2
Desk top stencil scanner (Gestetner E.S.455.)

PLATE 3
Typewriter with changeable founts (Varityper 720).

PLATE 4

Rotary offset printer (Rotaprint R18/75).

PLATE 5
Fount segment for Varityper.

PLATE 6
Electrostatic photocopier (Xerox 914).

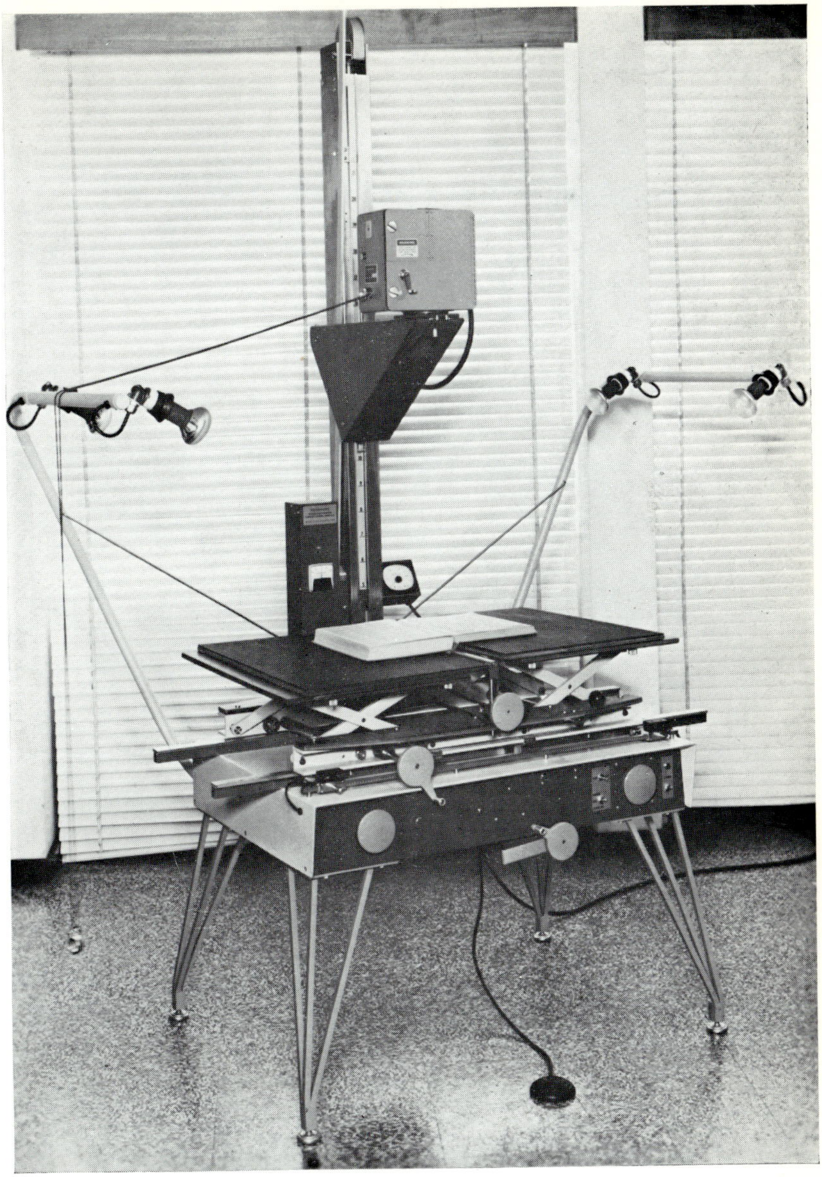

PLATE 7

Planetary microfilm camera (Recordak Microfile MRD-2LE).

Plate 8
Portable rotary microfilm camera (Recordak Microfilmer RP-1).

Plate 9
Microfilm reader (Recordak Lodestar).

PLATE 10
Microfilm reader/printer (Recordak Magnaprint).

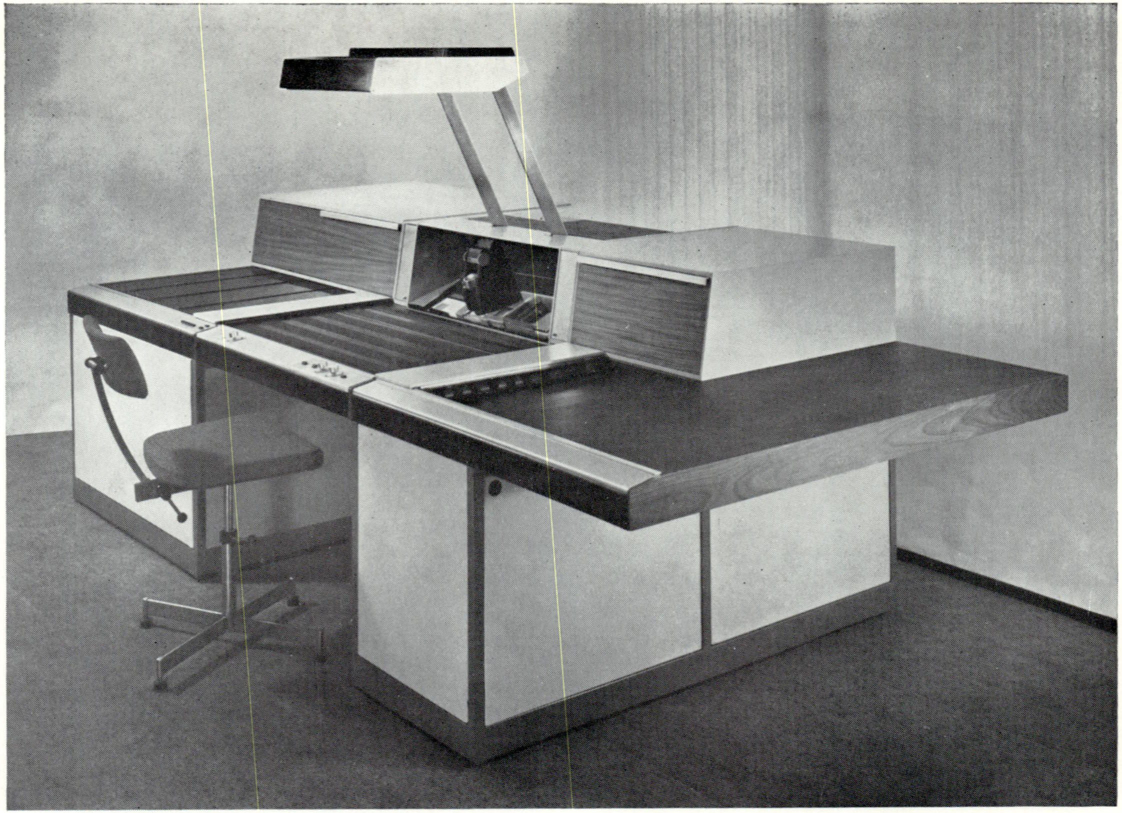

PLATE 11

Microfilm enlarger/printer (Caps M9 production microfilm printer).

printed catalogue is to be prepared from one which exists in card or sheaf form, and where an existing printed catalogue needs to be brought up to date by addition of entries from other sources. In the first case the cards or sheaves can be copied directly on to offset-litho plates by xerography or a photocopying process, or on to photographic stencils. However the average sheaf takes up a lot of room on a page, and 5 in. × 3 in. cards do not fit easily into normal sizes of paper so this method will only give a few entries to the page. There is therefore a need to reduce the size of the entries, usually about 20%, because this will allow the contents of two 5 in. × 3 in. cards to be put across the width of 8 in. wide paper, thus producing two columns of entries per page. This can be done by locking the cards into a glass frame and photographing through a reducing camera. The image can be put on to an offset plate via xerography or onto a photographic stencil. Alternatively, the same format can be used but a greater initial reduction achieved by using a microfilm camera. After development the microfilm is fed through a Xerox *Copyflo* and enlarged xerographic copies are made directly on to a roll of offset-litho paper masters. The roll is then cut up into page lengths and used to print the catalogue.

The obvious example of the preparation of a new printed catalogue from an old one brought up to date with later entries is the new British Museum *General catalogue of printed books*. The previous edition of the catalogue had begun in 1931, but twenty-three years later only fifty-one volumes had been issued and these only covered the letters A to DEZ in the alphabetical sequence. The difficulties lay in preparation and in proof-reading, particularly in the former, because considerable editorial work was involved in sorting out a pasted up sheaf catalogue in which the later entries had first been stuck in a blank column alongside entries from the original printed catalogue, and later as the columns filled up, stuck in the nearest available place. To overcome these problems a special Abstractor camera was designed, by the Williamson Manufacturing Co., which could cover a variable field and was adjustable so that it could photograph single entries from one line to fifteen lines in depth. The entries in the sheaf catalogue were photographed in their correct order on to 4 in.

wide rolls of silver halide translucent paper. The camera optics were designed to give a same-size copy. After development the translucent paper negatives were used to prepare special double sided pre-sensitised offset-litho plates, which in turn were used to print the new catalogue. To complete this large project as quickly as possible a production target of one new volume per week was set, and the organisation for this required an editorial team to number the entries in the correct sequence, and a trained staff to operate six Abstractor cameras in order to photograph the requisite 60,000 entries per month.

Information bulletins

The information included in current awareness bulletins is often also recorded in another form in the library. Abstracts of patents and technical literature are put on to index cards and filed, and recent additions of books, pamphlets, trade literature, etc. are catalogued and/or indexed. Once again the emphasis in using document reproduction methods lies in the preparation of both the bulletins and the other records with only one typing and one proof-reading. This can be achieved in two ways. The first is by preparing page masters, printing the bulletin, and then cutting up the masters and printing cards from the portions. This can be done fairly easily with offset-litho plates and spirit duplication masters, but is a somewhat messy business with wax stencils. The other method is by printing first the index and/or catalogue cards and during this process printing one extra white card for each item. These white cards are then stuck on to sheets of paper, overlapping where necessary to cover up the unused parts of the cards. A two column format is best with 5 in. × 3 in. cards stuck side by side on to 10 in. wide paper. Pages made up in this form can then be reduced by a camera on to the normal 8 in. wide printing master, which can be a photographic stencil or a sensitised offset-litho plate. Alternatively the reduced image can be put down by xerography on to an offset-litho plate. The master is then used to print the pages of the bulletin.

Information storage and retrieval

Basically an information retrieval system acts like a series of sieves. The first sieve is the index, in whatever form it may be, and the approach via this tool allows the enquirer to screen out the major part of the unwanted information contained in the store. The selected index term entries sometimes lead directly to documents but more frequently to abstracts. The abstracts, by providing fuller information than the index entry, again act as a sieve and allow the enquirer to reject those which are not completely relevant to his purpose. Finally a serial number or some other reference gives the address or location at which the original document can be found. The purpose of the use of microforms in information retrieval systems is to either remove one of these steps by providing an abstract in microform alongside the index entry, or better still to remove both steps by providing the complete document in microform. This is a developing field in which a lot of experimental work is taking place. The following descriptions are included to show how some of the various systems operate.

The idea of a microfilm rapid selector was originally conceived by Dr. Vannevar Bush before the last war. The design, construction and testing of the first machine were supervised by Ralph R. Shaw, when Librarian of the U.S. Department of Agriculture.

The material, usually abstracts, is recorded on to 35 mm. film in such a way that the image appears only on one half of each frame. On to the other half is photographed a code designating the subjects covered in the abstract. Thus each film is divided into half down its length, each half frame on one side having a micro-image of an abstract, and each half frame on the other side having the related code.

The code consists of twelve horizontal rows of black and white squares, each subject being designated by a different pattern. When a search is projected an interrogating card is made, using a special punch which makes an inverse pattern (on the card) to the pattern for which a search is being made.

The interrogation card is inserted in the machine and the search can then begin. The microfilm is run through the selector at high speed (180 frames per second) and the code part of the film is projected on to the interrogation card. Behind the card is a group of photo-electric cells which receive the light passing through the holes in the card. When a code pattern exactly matches the pattern on the interrogation card, the result is a 'black-out' for the photo-cells. This puts into action the copying mechanism which makes a two-times enlargement of the abstract portion of the frame on to another microfilm. All this takes place within a fraction of a second and the film continues as before without stopping.

When the search is over, all the relevant items will have been copied on to a piece of microfilm. This can then be either read in a microfilm reader or else enlarged prints can be made.

The storage capacity can be measured in units of 2,000 feet reels of film, each carrying 72,000 abstracts. Each reel can be searched within six minutes.

Disadvantages are the possibility of the photo-electric cells breaking down, and the wear and tear on the microfilms.

The *Recordak Miracode* system uses the same basic idea as the microfilm rapid selector. A special flatbed 16 mm. camera is used to make the film. The documents to be recorded are placed on the platen and the index codes are set manually by the operator by using a switch keyboard. The codes are then exposed automatically on to the film adjacent to the document concerned. After processing the 16 mm. film is loaded into cartridges. Searching is carried out on a special version of the *Lodestar* reader/printer using a keyboard control unit which can be set for searching either negative or positive microfilm. After the search parameters have been set up, the operator presses the search button and the desired document appears on the screen in an average time of 8 seconds.

The Eastman Kodak Company have developed a system of recording images and codes on to pieces of microfilm known as *Minicards* (see figure 12). Each piece is 16 mm. × 32 mm. in size and can copy up to twelve images photographed at a reduction of 1/60. The *Minicard* combines the manipulative abilities of the punched card with the high-storage capacity of microfilm.

The camera which makes *Minicards* has a two-step exposure. The first is for the code pattern which is taken from previously prepared punched paper tape. After all the code has been recorded the second exposure photographs the document pages. The film in the camera is in roll form and is cut into separate *Minicards* after processing.

Machines have been developed for both the sorting and the selection of *Minicards*. The sorter works on the same principle as the punched card sorter and directs the *Minicards* into the appropriate storage magazine. In the selector, *Minicards* are fed past a reading head which scans the code field. Those carrying the code

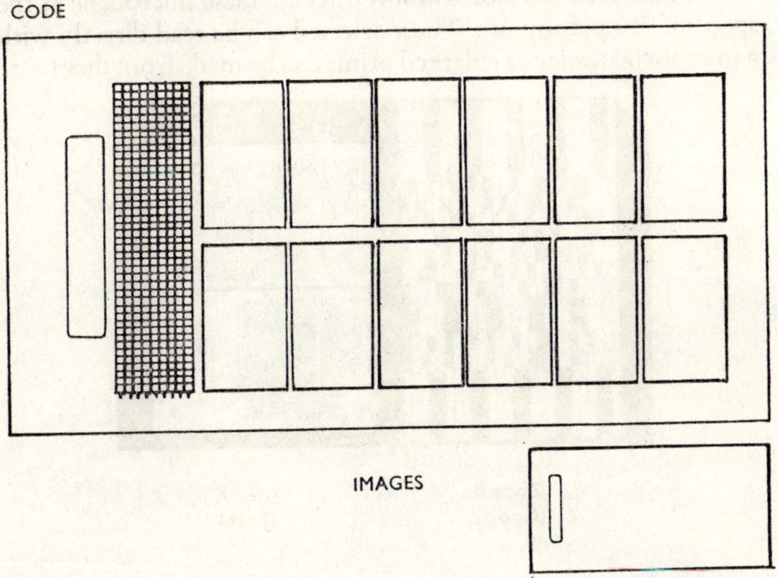

Figure 12.

data for which the machine is searching are passed into a special magazine. After the search the *Minicards* selected can be examined in a reader, and the enlarged prints can be made if required.

Minicards are stored in standard magazines, each of which holds 2000. These magazines can be combined into 10-magazine or 100-magazine file-units. A file of 2,000,000 *Minicards*, roughly equivalent to 1000 normal filing cabinets full of documents, will occupy a cabinet about 15 in. × 30 in. × 50 in. in size.

A similar system, known as *Filmorex* (see figure 13) was devised by Dr. Jacques Semain, Chef de Service of the Centre National de la Recherche Scientifique, Paris. A camera with two lenses is used to photograph the code and the documents on to film 72 mm. in width. Each frame is 45 mm. long. After the film has been exposed and developed it is cut into its individual frames. Thus each document and its code is contained on a microfiche 72 mm. × 45 mm.

The Filmorex Selector will sort through these microfiche at the speed of 600 per minute. Those selected can be read directly with a microfiche reader or enlarged prints can be made from them.

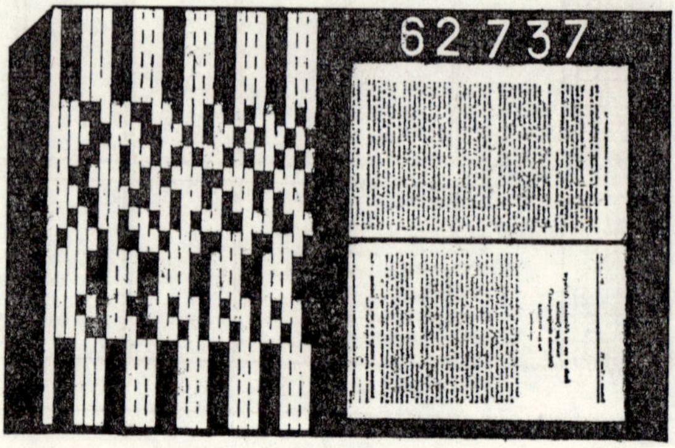

Zone B (Code) — Zone A (Text)

Filmorex microfiche

Figure 13.

An Aperture card is basically a machine sorted punched card in which a square or rectangular hole has been cut into which a microtransparency can be easily inserted. The body of the card is punched with sorting holes which represent in code the subjects contained in the original text, and the insert contains the text in microform. A sorting machine is set up to search for the pattern of holes which represent the desired subject combination, and the aperture cards are passed through it. As each card passes the sorting head, electrical contact is made through the holes punched in the card and the required cards are passed on into a sorting position, all the others being dropped into the rejected position. Care has to be taken, with some sorting machines, to ensure the electrical contact brushes are removed from that part of the sorting head corresponding to the apertures on the cards, otherwise the microform can get scratched. The rejected cards can be returned to the filing trays, while the accepted cards are taken to a microform reader/printer. The contents of each microform are read on the screen, and an enlarged print made from those which are required for further consultation. The major use of aperture cards at the moment is in the handling of engineering drawings. Such drawings are bulky, not easy to handle, and soon get out of date. It has been found that a large proportion of plant erection costs can originate from alterations made necessary because the original work had been done from an out-of-date drawing. Thus a system in which all drawings are held on aperture cards which are up-dated frequently as the drawings are altered and corrected, and in which all drawings are re-issued each month in a different coloured card to allow an easy check, can more than pay for itself.

Optical coincidence cards, sometimes known as Batten cards or *Peek-a-boo* cards, are used in a form of co-ordinate index. Each card represents a subject term or keyword, and has on it a series of numbered positions each one of which represents a document. If any of the documents includes information on that particular subject, then the position representing the document has a hole punched through it. When an enquirer consults a file of these cards he extracts the cards bearing the subject terms in which he is interested. He puts the cards together in a deck and he holds them up to the light, and makes a note of the numbers of the positions where there are holes right through the deck, for these represent those documents which contain information on all the subjects for which he is searching. Next he turns to a numbered series of abstracts and looks up the numbers which he has noted. After reading the abstracts he decides which are the documents he wishes to see and obtains them from the document store. A microform system known as *Microcite* has been developed which removes one of these steps and thus simplifies the procedure. A sheet of film is prepared which corresponds in size to the optical coincidence cards, and which carries in equivalent places to the numbered positions on the cards micro-images of abstracts of the documents represented by the positions. Thus when the selected cards have been put together in a deck, the film sheet is placed with them and the deck inserted into a special microform reader/printer. The holes which pass through the entire deck are placed one at a time in the reading position, and the abstract of the document appears on the screen of the reader. Enlarged prints can be made of those required for further consultation.

A collection of trade catalogues can take up a lot of space in a library and unless properly indexed can be fairly intractable material. The cost in storage space plus the cost of indexing can be fairly high, therefore anything which reduces these must be of interest. A system has been developed in which the pages of the trade catalogues are microfilmed and given a numerical sequence which is coded on to the film. The contents of the catalogues are indexed using optical coincidence cards. When a search is made the cards are consulted in the normal way and the numbers of the holes noted. The appropriate microfilm cartridge is removed from the rack and placed in the reading and sorting device. The first number is fed to the sorting input and the microfilm passes rapidly through the reader and stops at the required page. This process is repeated until all the pages on that particular microfilm have been consulted. A rapid re-wind returns the microfilm to the cartridge, the cartridge is replaced in the rack and the next cartridge placed in the reader/sorter. Systems of this type are becoming commercially available and although their annual rental is not cheap, the savings effected in space and indexing time should more than compensate.

Administrative uses

The possession of a duplicating or offset-litho printing machine allows the library to prepare and print its own forms and thus save some of the printing costs. Readers application forms, tickets, overdue notices, photocopying application forms, reservation forms, and all the usual array of library stationery can be organised on a 'do-it-yourself' basis. A note of warning must however be struck at this point. Due consideration must be given to the correct design of the forms. There may be a temptation to put everything possible on to one gigantic multi-part form. This must be resisted.

Each form should be for one purpose only. It should not ask for any unnecessary information. Its arrangement should follow a logical order, and the instructions for filling it in should be as simple and as easy to understand as possible.

Other items which can be produced locally are notices of meetings and other activities; addressed envelopes for circulation of notices, agenda and minutes of committee meetings; and staff instruction manuals.

In most libraries the staff record files will probably not take up a lot of space, but in those where they do consideration can be given to turning them into microform. If the entire series of files are to be converted, a method should be used that will permit the addition of new material. The contents of each file can be put on to 16 mm. microfilm and the film cut up and put into microfilm jackets, with one jacket per member of staff. Additions to the files can be microfilmed in batches, then cut up and inserted into the appropriate jackets. Alternatively microfilm can be used to make micro-opaque strips which can be stuck on cards to form micro-cards. It may however be simpler to keep the files of staff still employed at the library in the normal form and save space by microfilming only the completed files of those staff who have left. These can then be kept as microfilm rolls and a simple index will be sufficient for the occasional reference to them.

The use of microform as an issue method is well known, so only a brief description is given here. The charging device is a micro-film camera, and as each book is issued the book number, the serially numbered transaction card, and the reader's ticket are photographed on to one frame. It is often possible to photograph two such records on to a frame. The reader keeps his ticket and the transaction card is filed in a pocket in the book. When the book is returned the transaction card is removed before the book is replaced on the shelves. On the date chosen for the sending out of overdue notices, the transaction cards for the overdue period concerned are sorted into order and the missing numbers are noted. A search is then made through the microfilm for that period and the details of the book and the reader are ascertained. The process can be further mechanised if the transaction cards are punched with their serial numbers so that they can be automatically put into order by a punched card sorting machine. The use of a reader/printer for consulting the microfilm also permits the issue record to be printed out, thus saving the transcription of the details on to overdue notices.

The handling of rare and expensive books and manuscripts should be avoided as much as possible, and while there will always be some scholars who need to consult the originals, in many cases a copy will suffice. Microform copies are best because in the making of these the least damage is likely to be done to the original material. One other aspect of security copying needs mention. Many special libraries have collections of confidential reports which represent in effect the corporate memory of the library's parent body. In order to avoid complete destruction of these records by fire, they should be micro-copied, and one or more sets of the microforms stored in a safe place on another site.

The question of how to put a quart into a pint pot is one of the major problems of librarianship. Storage costs are high, binding costs are getting higher, and all the time more and more books are being published and periodical files get larger and larger. One of the solutions put forward is the use of microcopies to reduce the amount of storage space needed. This is particularly applicable to back files of periodicals and newspapers. In many cases heavy use is made of periodicals when first published, and then the use reduces until, after a period of five to ten years, any particular volume is consulted very infrequently. The amount of storage space taken up by the files is not justified and yet the occasional use suggests that they should not be discarded. Microcopies can be made or purchased for little more than the cost of binding the originals.

The amount by which the storage space can be reduced differs according to the microcopying process and the method of storage used. For example a file of periodicals occupying 10 cubic feet will reduce in 35 mm. microfilm form to approximately 0.45 of a cubic foot, and in microfiche or microcard form to approximately 0.25 of a cubic foot. This will obviously produce a considerable saving in storage costs, but these are not the only costs to be considered. On the one hand periodicals stored as microcopies can save the cost of binding. On the other hand there is the cost of the microcopies and the cost of the necessary equipment such as readers, reader/printers and cameras. When all the costs have been assessed it will be found that there is only a small overall saving in using microforms. This fact will be relatively unimportant if the library is so packed full that only a microform programme will ease the situation, because in this case the costs of the microforms and equipment have to be compared with the cost of a new building. However if the library still has storage space available, then it is my belief that it is the librarian's duty to keep, for as long as possible, the original material for his readers to use. It is also necessary to be aware of other methods for reducing storage needs. Little used material can be kept on a regional basis, either in joint deposit libraries, or on the basis of a subject specialisation scheme. Also files of periodicals which can be obtained easily on loan from the National Lending Library for Science and Technology, can be discarded once the initial period of heavy usage is over. However microcopies and co-operative storage should not be considered as alternatives, but as complementary systems.

CHAPTER 5

Bibliographical Control

Where reproductions of documents are made, purchased, catalogued, and stored, various forms of bibliographical control are essential in order to ensure that the different items are correctly labelled and recorded for identification and retrieval purposes. The major part of this chapter will be taken up with the control of microforms. The other types of document reproduction present fewer problems and are therefore dealt with first.

Printed and duplicated documents

The information to be included as part of these documents will depend on the use to which the document is to be put. Reading lists, bibliographies, printed catalogues, etc., will all require full details of the title, compiler, name of library, place of publication, date, and series number (if part of a series). This information can be printed on the cover or on a separate title page. The pages which make up the document should be correctly numbered in a prominent place. Apart from giving ease of reference for the user, this will also make for ease of collation before binding or stapling. The cataloguing of such material is fairly straightforward because they can be treated as books or pamphlets.

Information bulletins issued by the library should have a bold title, volume number and part number, date, name of editor or compiler, and name of library. If the bulletin has a cover this information can be printed there, otherwise it goes as a heading filling approximately the top third of the first page. The pages are best serial numbered through the whole of each year's issues, and the items of information should be given serial numbers also, e.g. 67/1, 67/2 etc. This makes reference to individual items very easy for an enquirer. Bulletins, if kept and filed, should be catalogued as periodicals.

Forms for use in the library should be given a serial number when first designed and printed. This will facilitate reference and re-printing.

Photocopies

Many photocopies will be made for purposes of private study and research, and it is only necessary to check whether the bibliographical identification is included in the actual copies. Periodicals often have the title, volume number and date printed on each or on alternate pages, and, of course, the first page of the article will give the title and author. If any or all of this information is not part of the copy, the missing items should be written or typed at the top of the first page. Alternatively, as a matter of automatic control, the reader's application form, completed with the full bibliographical details by the library staff, can be photocopied on to a separate sheet, and this sheet attached on top of the completed set of photocopies so that it becomes, in effect a title page. It will also serve as a reminder to the reader of the conditions under which he acquired the copy.

Photocopies made for addition to the library stock should have full bibliographic details on them, and should be catalogued as though they were the originals. The words 'photo-copy' or 'enlarged prints from microform' should be added to the catalogue entry. Such photocopies can present problems in storage and filing particularly if they consist of only a few pages. It may prove to be best if they are stapled into light cases and filed in a separate series with a special serial number for identification. They can be stored in vertical or lateral files and easily found via their entries in the catalogues or indexes. Photocopies of complete books should be bound and placed on the shelves as though they were the original copy.

Microcards and Microfiche

These two types of microform are essentially the same and are therefore dealt with together and first. The bibliographical information to be included on the microform itself includes certain essential items. Firstly a short title which can be read without the need of a magnifier. Any serial number should be given in the top left hand corner. In microcopies of books and pamphlets, author, title and date should be given. Periodicals should have title, date, volume number, part number(s) if necessary, inclusive page numbers, and the word 'index' if the index pages are included on that particular card or fiche.

Secondly a bibliographic description of the original which should be the first micro-image(s). This should include the author, title, publisher, date (or period covered), number of volumes, pages, place of publication, editor or translator, etc., of the original material copied. If only a part of a publication has been copied, then the description should relate only to that part, but the reference of the whole publication should also be given in brackets. Other details which must be given are the date at which the microcopy was made; the name of the microform publisher, library or other agency where the microcopy was made, and the name of the distributor if different; the location of the master negative; and the reduction ratio of the micro-images. In some cases it may be necessary to give a detailed list of the contents, or a list of gaps which were present in the original material. If rare or unique material is copied, the location of the original should be given, and manuscripts may well require an introduction. Finally if there are any restrictions on the use of the microforms, a clear and unambiguous statement of such restrictions should be given.

The arrangement of micro-images on the microcard or microfiche should be first the bibliographic details recorded in the upper left hand frame, and then the title page and succeeding pages should be recorded in numerical sequence from left to right, then downwards from row to row. The last frame of text should be followed by a frame carrying either CONTINUED or END as necessary. If a resolution test chart is included it goes in the last frame, i.e. the frame in the bottom right-hand corner.

The catalogue entries for material held as microcard or microfiche should be exactly as for the original material, with added notes on the particular microcopy held. Such notes should give details of the form of the microcopy; its size, and the reduction ratio of the micro-images; the number of individual microcards or microfiche; and in the case of microfiche, whether negative or positive. As in the bibliographic description it may also be necessary to give the date and name of maker of the microcopy, the location of the master negative, and the location of the original material.

Until the time when these two types of microform are made in one standard size, it will be necessary to file them first according to size, in the appropriate size of index drawer. The exceptions to this at the moment are those sets of *Microprints* which are supplied complete with covers so that they can stand on shelves as though they were books. There does not seem to be any good reason why microcards and microfiche of the same size should not be interfiled, and to do so is economical of drawer space. The internal arrangement in the index drawers can be alphabetical, accession number order, by form of original (book, periodical, etc.), or a classified order. The choice of arrangement will depend on whether the collections of microforms are centralised or de-centralised, and whether the library users are permitted to find and remove items from the drawers or whether staff only are allowed to handle them. Probably the best arrangement is firstly by form of original, thus separating the collections into books and pamphlets, periodicals, patents, etc.; and secondly books, etc. by author, periodicals by title, patents by serial number, and any other material by some equally important form of reference. Such an arrangement permits quick retrieval from the original reference and avoids the need to use the catalogue or index when the existence of the microform in the library is already known.

Microfilm

The length of a full reel of 35 mm. microfilm is usually 100 feet and the number of micro-images on the reel can be as high as 1,500. This calls for careful bibliographic control, otherwise it becomes very difficult to find the required item on the reel. There are automatic and semi-automatic locating devices now available but even so it is best to continue to provide full control.

The first item on the microfilm should be the identification short title, readable without the aid of a magnifier. If there are several different documents photographed on the film, then all should be listed in this easily readable form and given serial numbers and/or the inclusive numbers of the frames which they occupy. The word START should also be included in a prominent position. The second item on the film can be, if desired, a resolution test chart, otherwise the next item is SERIAL NUMBER 1 followed by, on the next frame, the full bibliographic description of the first document on the film. This description should include as essential the following: Author, Title, Publisher, Date or period covered, Number of Volumes (or pages), Place of publication. Edition, Name of microfilm publisher or the library which made the microfilm, Date of making, and Reduction ratio. If desired, information can also be included on location of original document, location of master negative, and any restrictions on use. If any unusual system of image placement has been used, then this should also be mentioned.

In a very few cases the item microfilmed occupies more than one reel, in which case the next frame after the full bibliographic description should contain a list of the items on that particular reel. Otherwise the next frame should be a micro-image of the words FIRST PAGE OF TEXT. After this comes the title and then the contents pages of the first document, followed by the index pages and then by the text. It is important to move the index to this position in order to save time for the enquirers who wish to locate a piece of information rather than read through the document. At the end of the text there should be a frame with the words LAST PAGE OF TEXT, followed by a repeat of the identification short title, but this time with the word END in a prominent position. A gap of six frames should be left before the next item, so that the seventh frame will be SERIAL NUMBER 2, and will be followed by the bibliographic description of this document, and so on. The identification short title of the final item should carry the words END OF FILM, PLEASE RE-WIND.

With several documents on one reel, the above method makes it fairly easy to find a particular document. As the film is passed through the microfilm reader, the unused frames between each item shew up quite well on the screen as blackouts on positive film and white flashes on negative film. These can either be counted, or the film can be stopped at this point and the next frame will give the serial number of the next document. Only a slight modification is needed to deal with a document that occupies the whole reel. Every 50 pages a frame is used to photograph, in enlarged figures, the number of the next page. The figures actually photographed are white and are placed against a black or chequered background. These will shew as flashes when the reel is being wound through the microfilm reader, and the film can be stopped for the number to be read. Most documents microfilmed will have their own page numbers, and these will provide the control necessary to identify a particular page. If for any reason such numbers are not in the original, then they must be provided. The easiest way to do this is to type or letter the numbers on to white paper, cut them out and lay them on the page when it is microfilmed.

Microfilm is catalogued as though it were the original document, with the usual added note covering type and size of microfilm, number(s) of reels and position on reel where necessary, reduction ratio, negative or positive, date made and name of library or publisher, location of original, and location of master negative.

Reels of microfilm are kept in cardboard boxes, and there are special filing cabinets for their storage. Boxes holding 35 mm. and 16 mm. reels vary only in thickness so can be stored together if so desired. The simplest method is to file in accession number order and see that the reel is marked and that all records specify this number. The accession number should also be included in the identification short titles on the film as a safeguard against the film being re-wound on to the wrong reel.

Short lengths of microfilm on a full sized reel are often added to the stock of a library and can cause quite a loss of storage space if filed individually. Provided it is possible to include sufficient bibliographical control to make each item easily found, probably the best method of dealing with them is to splice a sufficient number together to occupy a full reel. Alternatively the film can be cut into strips, filed in transparent jackets and treated as though microfiche.

Microforms have particular problems of identification and location, and require bibliographical control through lists of published microforms, union-catalogues and lists of microforms held in libraries, and location lists of master-negatives. Published microforms are recorded in the various publishers lists, e.g. the *Microcard microfiche catalog* of Microcard Editions Inc., the *Microfilm catalogue* of University Microfilms Inc., and the general *Catalogue* of Micro-Methods Ltd. University Microfilms also issue their *OP catalogue* of out of print books which have now been microfilmed and are available for purchase in the form of bound enlargements made by the Xerox process. The *Guide to microforms in print* published by Microcard Editions Inc., is a list of microform publications available for purchase. Books are entered under the author's name, periodicals and sets are entered under the title, and newspapers under the location. There is also a *Subject guide to microforms in print* issued by the same publisher. The catalogues and lists quoted are only a few of those now available, and as the microform comes to be more and more accepted as a method of publication, the number will no doubt increase to the stage where librarians will receive as much publisher's literature on microforms as they do at present on books.

Other types of material are covered by special bibliographies, for example the reports issued by certain U.S. Government departments are listed in the twice-monthly publication *U.S. Government Research and Development Reports* and can be purchased in microform; and the University Microfilms periodical *Dissertation Abstracts* lists and abstracts those theses from American Universities which are available as microfilms. A number of theses from British Universities are included in the Micro Methods Ltd. *Catalogue*, and they also issue a separate *Catalogue of U.K. Atomic Energy Authority Unclassified Reports*.

Union lists of library holdings of microforms and master negatives have been much slower to appear. Such lists are useful both in providing locations of required material, but also in avoiding the duplication of microcopying work already done. Most of the work in the field has been done in the United States, and includes the Philadelphia Bibliographical Center Union *List of microfilms*; and the Library of Congress *Newspapers on micro-*

film and *Check List of the microfilms prepared in England and Wales for the American Council of Learned Societies*. The Library of Congress with the aid of a grant from the Council on Library Resources is building up a *National Register of Microform Master Negatives*, which will record the location of master negatives of books, newspapers and serials and publish this informaton in a series of reports. The Library Association published a list of British *Newspapers on Microfilm* in the Library Association Record, August 1960, which gave under each title the years covered, whether complete or not, the holder of the master negative, and the cost of a positive microfilm copy (if such copies were available for purchase). More work will, no doubt, be done in this field as the need for it is felt by a large number of librarians. In the meantime all microforms acquired by libraries should be included in the appropriate union catalogues, in the same form of entry as in the library's own catalogue, i.e. the entry should be as for the original document and the details of the microform should be given in an added note.

CHAPTER 6

Copyright

Until the development of photocopying, students had to take notes and make extracts by either writing or typewriting. Both these processes were sufficiently laborious to prevent any major infringement of copyright. Photocopying saves time spent in making notes, but, because of the ease with which copies can be made, publishers have felt that the sale of publications could be affected, particularly if photocopies were supplied at a cheap rate.

In 1951 the President of the Board of Trade appointed a committee whose terms of reference were 'To consider and report whether any, and if so what, changes are desirable in the law relating to copyright in literary, dramatic, musical and artistic works with particular regard to technical developments and to the revised International Convention for the Protection of Literary and Artistic Works signed at Brussels in June, 1948, and to consider and report on related matters'.

The Act in force at the time was the Copyright Act of 1911. This Act was based on the general provisions of the original Berne Convention of 1886 as revised at Berlin in 1908. However, since the passing of the 1911 Act there had been two International Conventions of the Berne Union and parts of the Act were outdated.

The Report of the Copyright Committee was presented to Parliament in October 1952 and, together with the Brussels Convention and the Universal Copyright Convention, became the basis of the Copyright Act, 1956.

The part of the 1956 Act which covers the making of copies is Section 7, entitled 'Special exceptions as respects libraries and archives'. The following is a summary of the relevant parts of that section, together with appropriate regulations laid down in the Copyright (Libraries) Regulations, 1957.

1. *Periodical publications*

The copyright in an article contained in a periodical publication is not infringed by the making or supplying of a copy, if it is made or supplied by or on behalf of the librarian of a library of a class prescribed by regulations made by the Board of Trade, and the conditions prescribed by those regulations are complied with.

The Board of Trade when making these regulations are to ensure:

(a) that the libraries are those which are not established or conducted for profit.

(b) that copies are supplied only for purposes of research or private study.

(c) that no person is supplied with two or more copies of the same article.

(d) that no copy extends to more than one article contained in any one publication.

(e) that the persons to whom the copies are supplied are required to pay for them a sum not less than the cost (including a contribution to the general expenses of the library) attributable to their production.

The Copyright (Libraries) Regulations, 1957, specify the following types of libraries for the purposes of the subsections relating to copying of articles from periodicals and also for literary, dramatic and musical works other than articles:-

1. Any library (including the British Museum) to which section 15 of the Copyright Act, 1911, applies.
2. Any library of a school, university, college, or any library of an establishment for further education.
3. Any library administered by a Library Authority under the Public Libraries Acts.

4. Any Parliamentary library or library of a Government Department.
5. Any library, conducted for, or administered by any establishment or organisation for the purpose of facilitating the study of all or any of the following: religion, philosophy, science, technology, medicine, history, literature, languages, education, bibliography, fine arts, music or law.

2. *Literary, dramatic or musical works other than articles*

Copyright is not infringed by making or supplying a copy of part of the work, if the copy is made or supplied by or on behalf of a librarian of a library of a class prescribed by regulations made by the Board of Trade, and the conditions prescribed by those regulations are complied with. This, however, does not apply if, at the time when the copy is made, the librarian knows the name and address of a person entitled to authorise the making of the copy, or could ascertain the name and address by reasonable enquiry.

The regulations for this subsection are to be the same as those applying to periodical articles except that paragraph (d) does not apply. In place of this the Board of Trade has to ensure that no copy is made which extends to more than a reasonable proportion of the work.

3. *Copies made by one librarian for another*

Copyright is not infringed if the copy is made or supplied by the librarian of a library of the prescribed class, providing

(a) the copy is supplied to a librarian of a library of the prescribed class.
(b) the librarian supplying the copy does not know the name and address of a person entitled to authorise the making of the copy, and could not by reasonable enquiry ascertain the name and address. (This paragraph does not apply to periodical articles.)
(c) any other conditions prescribed by regulations are complied with.

The regulation for this section prescribes all the types of libraries given at the end of section 1 above and also includes any library which makes works in its custody available to the public free of charge. This regulation also applies to any library of a class so specified whether established or conducted for profit or not.

4. *Manuscripts*

Where, at a time more than fifty years from the end of the calendar year in which the author of a literary, dramatic or musical work died, and more than one hundred years after the time at which the work was made:-

(a) copyright subsists in the work, but
(b) the work has not been published, and
(c) the manuscript or a copy of the work is kept in a library, museum or other institution where it is open to public inspection,

the copyright is not infringed by a person who reproduces the work for purposes of research or private study, or with a view to publication.

5. *Prescribed conditions*

The following conditions apply to the making of copies of periodical articles and copies of parts of other works:-

(a) no copy shall be made for or supplied to any person unless he has completed the following form of declaration and undertaking.

To

The Librarian of

the Library

(Address)

1. I[ ] of[ ] hereby request you to make and supply to me *[a copy of (particulars of article)] *[a copy of (particulars of work of which a part is required to be supplied and particulars of that part)], which I require for the purposes of research or private study.

2. I have not previously been supplied with a copy of *[the said article] *[the said part of the said work] by any librarian.

*Delete whichever is inappropriate.

3. I undertake that if a copy is supplied to me in compliance with the request made above, I will not use it except for the purposes of research or private study.

Signature* ....

Date ....

[**Note:* This must be the personal signature of the person making the request. A stamped or typewritten signature, or the signature of an agent is NOT sufficient.]

(b) No copy extending to more than one article in a publication shall be made.

(c) No copy extending to more than a reasonable proportion of a work shall be made.

(d) The charge for the copy shall not be less than the cost of production.

The following conditions apply to the making of copies by one librarian for another:-

(a) no copy shall be supplied to a librarian, if a copy of the same item has already been supplied to him, unless the librarian making the copy is satisfied that the copy already supplied has been lost, destroyed or damaged.

(b) librarians to whom copies are supplied must pay for them a sum not less than the cost of production.

(c) no copy shall be supplied to the librarian of a library that is established or conducted for profit.

Fair copying declaration

The problem of obtaining permission to make copies was discussed at length during the Royal Society Scientific Information Conference in 1948, and further considered by the Information Services Committee which was set up by the Royal Society to advise on the implementation of the Conference recommendations. The result of these deliberations was the drawing up of the 'Declaration on Fair Dealing in regard to copying from scientific periodicals', the text of which is as follows:-

For some time scientists have discussed the problems created by the Copyright Act, which arise when they wish to obtain reproductions of excerpts from scientific and technical periodical publications. In the normal course of their work, scientists occasionally require for frequent reference, copies of particular papers appearing in scientific periodicals which are not readily available to them. It is assumed that they take all reasonable steps to secure the original journals or separates of papers they require, either from the author or the publisher, but it is recognized that many requirements cannot be met from these sources. We have, therefore, agreed to make the following declaration to ensure that scientists have no undue difficulties in obtaining copies from libraries and other organizations supplying information. This declaration does not apply to books and other non-periodic or non-serial publications.

We will regard it as fair dealing for the purpose of private study or research when a non-profit-making organization, such as a library, archives office, museum or information service, owning or handling scientific or technical periodicals published by us makes and delivers a *single* reproduction of a part of an issue thereof to a person or his agent representing in writing that he desires such reproduction in lieu of a loan or manual transcription and that he requires it solely for the purpose of private study, research, criticism or review, and that he undertakes not to sell or reproduce for publication the copy supplied, provided:

1. The recipient of the copy is given notice that he is liable for infringement of copyright by misuse of the copy, and that it is illegal to use the copy for any further reproduction.
2. The organization making and furnishing the copy does so without profit to itself.
3. Proper acknowledgement is given to the publication from which the copy is made.
4. Not more than one copy of any one excerpt shall be furnished to any one person.

The exemption from liability of the library, archives office, museum or information service herein provided shall extend to every officer, agent or employee of such organization in the making and delivery of such reproduction when acting within the scope of his authority of employment. This exemption for the organization itself carries with it responsibility to see that employees caution those receiving copies against the misuse of material reproduced.

We reserve the right to take action against any person or organization copying or misusing for any purpose whatever the whole or part of a work published by us without abiding by the conditions laid down herein unless the person or organization has our special permission in respect of the item to be copied. We reserve the right to withdraw this declaration.

A third edition of this declaration was published in 1957 and had, listed with it, the names of 143 publishing bodies which have subscribed to the declaration.

As both Section 7 of the 1956 Act and the Fair copying declaration refer specifically to libraries belonging to non-profit-making organisations, the position of libraries belonging to industrial and commercial companies has been somewhat obscure. The majority of such libraries have continued to make photocopies and have justified this service by referring to Section 6 subsection (1) of the 1956 Act which states 'No fair dealing with a literary, dramatic or musical work for purposes of research or private study shall constitute an infringement of the copyright in the work'. Fair dealing in this context is taken as meaning that the photocopy must not be of a substantial part of the complete work. In order to remain well within this somewhat arbitrary limit, many special librarians have (a) restricted provision of photocopies of articles in periodicals to members of the company to which the library belongs; (b) complied strictly with the Board of Trade regulations as they apply to non-profit-making libraries, with the exception of making a charge; and (c) refused to make photocopies of pages in books.

The position was made both clearer and easier by the publication by the Society of Authors and the Publishers Association in 1965 of a pamphlet entitled *Photocopying and the law*, the contents of which are reproduced below by kind permission of the publishers. A general permission is given to librarians of all kinds, and not just those covered by Section 7 of the 1956 Act, to make a single copy from a copyright work of a single extract not exceeding 4,000 words, or a series of extracts (of which none exceeds 3,000 words) to a total of 8,000 words provided that in no case the total amount copied exceeds ten per cent of the whole work. This general permission refers only to British copyright books. However as Article 2(1) of the Universal Copyright Convention requires for foreign authors the same degree of protection as that given to British authors, it would appear to be acceptable to apply the provisions of the general permission to foreign books.

PHOTOCOPYING AND THE LAW

The right of libraries of various kinds, and of individuals, to make photocopies (including micro-photocopies) of copyright material is widely misinterpreted. In general, the public at large (and libraries in particular) have wider rights than is often supposed; on the other hand, usually through ignorance of the law, abuse tends to be widespread. The object of this statement is to set out the facts.

There are two main points to be borne in mind by those who wish to copy copyright material:-

(i) No infringement of copyright can arise unless a substantial part of the copyright work is involved;

(ii) Even where a substantial part is involved, such copying may be permissible under the Copyright Act, 1965, as fair dealing

(a) for the purpose of research or private study (Section 6(1)) or

(b) for the purposes of published criticism or review (Section 6(2)).

It will therefore be useful to consider these factors first.

'Substantial part'

Under Section 49(1) of the Act there can be no infringement of copyright in a particular work unless a 'substantial part' of it is taken. This is not always fully understood but is of prime importance since it may be taken in law as overriding any other limitations contained in the Copyright Act. The Act does not, however, define what a 'substantial part' is.

It may, however, reasonably be supposed that the 'substantiality' of any extracts from a copyright work must depend on (a) the length of the work as a whole, and (b) the importance of the extracts in relation to the work as a whole. A claim that the extract was not 'substantial' could hardly be successful, for example, if the extract included a summary of the work (or of a part of the work dealing with a specific subject) either in words or, say, in the form of a diagram, graph, statistical or mathematical table or a scientific formula, so that the core of the work (or any specific part of the work) was contained in the extract.

Fair dealing in relation to research or private study

Section 6(1) of the Act states that 'no fair dealing with a literary, dramatic, or musical work *for purposes of research or private study* shall constitute an infringement of the copyright in the work'. Legal precedent, however, has established that such 'fair dealing' must be exercised by the individual for himself, and not by any one person on behalf of another. It thus follows that while a student or researcher may make copies of extracts from a work (or even the whole of the work) for himself, he may not make such a copy for another, without, in effect, 'publishing' the work, and thus infringing its copyright.

Fair dealing in relation to published criticism or review

In the absence of any legal definition of the extent to which copyright works may be used without infringement the Society of Authors and the Publishers Association jointly agreed in 1958 following the introduction of the 1956 Copyright Act, that, while the limits of 'fair dealing' must vary according to the circumstances, it would be a convenience to all concerned if a general understanding could be established for normal practice. They therefore agreed that they (i.e., the authors and the publishers, together representing the copyright proprietors) would not regard it as 'un-fair' if *for purposes of criticism or review* a single extract up to 400 words, or a series of extracts (of which none exceeded 300 words) to a total of 800 words were taken from prose copyright works, or if any extracts up to a total of 40 lines were taken from a poem, provided that in no case this extract amounted to more than one quarter of the poem. This joint statement was issued in respect of the provisions of Section 6(2) which relates specifically to 'fair dealing . . . for purposes of criticism or review'. The Act requires that any such use shall be accompanied by sufficient acknowledgement.

Photocopying by libraries

The rights of libraries to make photocopies of copyright works are set out as special exceptions in Section 7 of the Act. The only libraries entitled to take advantage of these exceptional provisions are those which are prescribed by the Board of Trade under regulations made under this Section of the Act, known as the Copyright (Libraries) Regulations, 1957, which define the privileged libraries as those operating in schools or establishments of further education covered by Section 41 of the Education Act, 1944 (and by Section 39 of the Northern Ireland Education Act, 1947, and Section 143 of the Scottish Education Act, 1946), as well as any public library or government department library or any library which exists for 'the purpose of facilitating or encouraging the study of all or any of the following: religion, philosophy, science (including any natural or social science), technology, medicine, history, literature, languages, education, bibliography, fine arts, music or law'. Such libraries do not, however, have the absolute right to make copies without restriction for anyone who requires them. Moreover, *any library established or conducted for profit is explicitly excluded.*

The libraries covered by the Board of Trade Regulations by virtue of Section 7 of the Act are free to make *single* copies of:-

(1) *Articles from periodical publications*, provided no person is supplied with a copy of more than one article from any one publication. It is generally assumed that the right of these libraries to make single copies of articles from periodical publications includes the right to make copies of articles from publications such as Annual Reviews and Year Books.

(2) *Extracts from copyright books*, either in cases where the copyright owner has given permission, or where alternatively, in cases where it is impossible to trace the copyright owner, the librarian is satisfied that every attempt has been made to trace him. Many libraries include in the formal undertaking they require from those who wish to have photocopies an assurance, accompanied by written evidence, that either permission has been obtained from the copyright owner or that it has proved impossible to trace that copyright owner. Librarians should under-

stand, however, that there is nothing in the Act or Regulations to afford them protection if they make a copy of such a work without permission having been obtained where (to quote the Act) 'the librarian knows the name and address of a person entitled to authorise the making of the copy, or *could by reasonable inquiry* ascertain the name and address of such a person'.

In both cases (i.e., both in respect of articles from periodical publications *and* in respect of extracts from copyright books), the Regulations lay down that single copies may be made by the privileged libraries (and by no one else) *only* for those who require them 'for the purposes of research or private study' and who have 'not previously been supplied with a copy . . . by any librarian'. And the libraries are required to obtain from the person requesting such copies a formal declaration and undertaking in writing to this effect. Thus the use of photocopying machines, which is nowadays a feature of the service provided by many public and other libraries, is clearly liable to be in contravention of the Act (and the Regulations) in cases where the users of such machines are left free to make copies without undertaking or hindrance. Librarians should bear in mind that they remain responsible for machines operated on their premises. It also follows, of course, that the unrestricted use of similar machines in railway stations and other public places may expose the user to the risk of committing a breach of copyright.

The privileged libraries which are allowed to supply photocopies are required by the Regulations to charge 'not less than the cost (including a contribution to the general expenses of the library) attributable to their production'.

The Microfilm Association of Great Britain (formerly known as the Council for Micro-photography and Document Reproduction) hopes shortly to suggest, on the recommendation of its Copyright Sub-committee, a form of undertaking which will enable the privileged libraries to ensure when issuing photocopies that they comply with the law. Such a standard form of undertaking will have the additional advantage that the form issued by one library may be accepted by another. At present there is a variety of forms which may in some cases lead to delay or difficulty.

In cases where a student or research worker might require a

complete copy of work that is out of print, he should get in touch with one or other of the companies which supply Xerox-copies or microfilm copies under licence from the copyright proprietor.

Limits of photocopying

Apart from the special exceptions covered by Section 7, photocopying may be permissible as 'fair dealing . . . for purposes of research or private study' under Section 6(1). The 1958 joint statement by the authors and publishers does not cover fair dealing of this sort.

Proper regard for copyright is essential to the continued publication of works of all kinds (technical, scientific and educational no less than biography, travel and fiction), for if free copying so reduces sales of the original commodity that publication becomes uneconomic, in the last resort important publishing in many fields must wither away. Without some reasonable control the source material, upon which those who wish to make photocopies depend, will cease to exist. The need thus becomes apparent for some sensible measure of 'fair dealing . . . for purposes of research or private study'.

The Society of Authors and the Publishers Association have, therefore, now suggested limits which authors and publishers generally would not regard as unreasonable for photocopying purposes. They have agreed that they would not normally regard it as 'un-fair' if a single copy is made from a copyright work of a single extract not exceeding 4,000 words, or a series of extracts (of which none exceeds 3,000 words) to a total of 8,000 words, provided that in no case the total amount copied exceeds ten per cent of the whole work. Poems, essays and other short literary works must be regarded as whole works in themselves, and not as 'parts' of the volumes in which they appear.

It has to be made clear that this rule of thumb measure is not a matter of law, but is merely an indication by two responsible societies of what they, as representatives of the generality of copyright owners, would regard as reasonable. It is probable, however, that any Court of Law called upon to determine whether copyright had been infringed by photocopying would take account of this joint statement.

The copyright proprietors represented by the Society of Authors and the Publishers Association wish to make it clear, also, that this general permission is intended to apply to librarians of all kinds and not only those covered by Section 7 of the Act and the Regulations attached to it. So far as librarians covered by the Regulations are concerned, the intention of this joint statement is to relieve them of the necessity of seeking permission (within the limits set out) even where 'the librarians know the name and address of a person entitled to authorise the making of a copy'. In cases where copies are made by virtue of Section 7 and the Regulations, this 'general licence' is not, however, to be taken as removing the librarians' obligation to obtain from the person requiring a copy the prescribed undertaking that the single copy he requires is 'for the purposes of research or private study' and that he has 'not previously been supplied with a copy' by anyone else. No such undertaking is *required* when copying is done under Section 6(1) although librarians might still prefer to ask for a similar undertaking to ensure that the provisions of that Section are met.

Illustrations

Librarians are also often asked to provide photo-positives (or even 'slides') of illustrations from copyright works, including graphs, charts, technical drawings and other kinds of illustrated matter. In general, works of this kind are regarded as complete copyright works in themselves (i.e., as artistic as distinct from literary copyright works) so that while the photocopying of such works might be thought of as no more than a photocopying of an 'extract', in fact it is nothing of the sort.

On the other hand, it has to be remembered that a few years ago, in a less technological age, schoolmasters and university lecturers had access only to more rudimentary equipment than they have nowadays. The epidiascope was a common feature of the classroom or lecture hall and by using it the lecturer could readily cast upon the screen the image of any page he chose from any book he wished—including illustrations of all kinds. Since such projection did not involve any physical 'reproduction', the epidiascope was not an instrument capable of infringing copyright. The photocopying machine is. It would, however, be an unusual publisher who would regard the reproduction of a single illustration for classroom or lecture hall projection as infringing his copyright, so that in general it may be regarded as reasonable for librarians to make such single copies without reference to the copyright proprietor. In fairness, however, librarians should ensure that any such illustration is accompanied by suitable acknowledgement to the source so that the copyright proprietor is at least given credit for the work.

Those who used epidiascopes did not always own the books whose pages they projected on the screen, but often borrowed them from libraries. Limited photocopying of this sort would therefore seem to be no more than a recognition of a new projection technique. The Society of Authors and the Publishers Association have therefore included such photocopying within the limits which they would regard as reasonable, provided its use falls within the limits laid down in Section 41 of the Act. This means, in brief, that the authors (and artists) and the publishers would not in general object to the making of a single copy of such an illustration (in the form of a slide or otherwise) provided it was used solely in the course of instruction in a school or other educational establishment.

This does not mean, of course, that any librarian (or anyone else) would be free to make *multiple* copies of illustrations to be handed round for class use. Such reproduction would be a clear infringement of copyright and permission must always be sought.

Multiple copies

We have this far been concerned only with the reproduction of single copies. It will be recalled that those libraries which are privileged under the Act are allowed to supply no more than a single copy of any photocopied article from a periodical publication or an extract from a book to any one person, provided that that person has not been supplied with a copy of it by any other librarian. Moreover, any single copy so supplied must not be used 'except for the purposes of research or private study'. The use made, both in the Act and in the Copyright (Libraries) Regulations, of the expression 'private study', read together with the special provisions regarding educational establishments (in

Section 41), makes it clear that it is not the intention of the Act that, for example, a number of students in a college or school should each obtain single copies of extracts which they may require for their communal studies in class; for such studies could not generally be regarded as 'private'.

Librarians in various kinds of libraries and educational establishments are nevertheless frequently asked by their teaching colleagues to produce multiple copies of extracts from copyright works for class or lecture hall use. Teachers should understand that when they make such a request the librarian is required by law to seek the copyright proprietor's permission. To ignore the law is to attempt (no doubt unwittingly) to undermine the economic basis upon which the publication of books for succeeding generations must depend.

The restrictions on the free use of copyright material for educational purposes are laid down in Section 41 of the Act. Under this Section, copyright cannot be infringed by the reason that the work is reproduced or adapted (1) in the course of instruction, whether at a school or elsewhere, where the reproduction or adaptation is made by a teacher or pupil otherwise than by the use of a duplicating process, or (2) as part of the questions to be answered in an examination, or in an answer to such a question. 'Duplicating process' includes virtually any kind of process involving the use of an appliance for reproducing multiple copies. It does not refer to the writing of a single copy by hand. School libraries *as such* are in general subject to the provisions of Section 7 of the Act and the Copyright (Libraries) Regulations, 1957, and are therefore intended to benefit from the 'general licence' to make single copies for research or private study within the limits set out in this statement. Multiple copies for class use, as explained here, do not qualify under this 'licence'.

There may sometimes be occasions when the reproduction of, say, two dozen copies of an extract from a particular work will not endanger the economic viability of the work at all. The law is there to ensure that the risk is avoided, but it would be unusual for a publisher to refuse to give permission for such copies to be made in circumstances where no such danger existed. The important thing is that such multiple copies may not be made without permission, so that permission must always be sought.

Conclusion

It is not possible within the limits of this statement to attempt to explore exhaustively all the possible uses to which the photocopying of copyright works can be put. It is hoped, however, that enough will have been said to indicate the general limitations of the law on the one hand, and of practicality on the other, which govern such use. Authors and publishers have made it clear by their 'general licence' that they do not wish unreasonably to restrict the photocopying of their works; but copyright exists for the encouragement of learning in the interests of society as a whole and must therefore be respected. It is hoped that this article has shown that the law, as interpreted by those who have practical experience of its operation, is not unduly restrictive, but that, at the same time, the unlicensed photocopying of copyright works must have some limit.

CHAPTER 7

The Document Reproduction Unit

Selection of equipment

This section is concerned with the principles of selection. It does not set out to present a *Which* report on document reproduction, for the simple reason that such a report would be out of date before it was in print. Library requirements have been considered in the three chapters dealing with processes, but it will not hurt to re-emphasise them at this point in order that the prospective purchaser of equipment has them clearly in mind.

For office printing and duplicating the average library requires moderately priced equipment that will produce prints on octavo, quarto and foolscap paper (or International paper sizes A4, A5, A6) and possibly also on 5 in. × 3 in. cards. The quality of printing must be good, and the print for some purposes (e.g. cataloguing) must be of a permanent nature.

A photocopying process must be able to produce cheap and reasonably permanent single copies from opaque or double-sided originals and be able to cope with large and/or thick bound volumes. The process should be quick and clean, should not need a skilled operator, and the equipment should be reasonable in cost, occupy as little space as possible and not need darkroom conditions.

Microcopies, if produced by the library, will need cameras, readers and reader/printers. If the programme is large enough, there may be a need for processing equipment, but usually it is better to have this side of the work done by commercial organizations.

The main question to be resolved is what do you want to do?

What types and level of services are to be provided for the library's clientele? What administrative and processing routines are to be simplified and improved by the use of document reproduction equipment? An account of the uses to which the document reproduction can be put has already been given in Chapter 4, and can be used as a guide.

When this decision has been made, examine the various methods which can be used to fulfil your requirements, arrange to see demonstrations, and choose the ones that will satisfy all your requirements with the smallest number of different processes. The variety of equipment in the Document Reproduction Unit can be kept to a minimum by the use of processes that can be used in conjunction with one another. Examples are the production of translucent masters by the chemical transfer process in order to produce multiple diazo prints; and the printing of one extra copy of each catalogue card and the mounting of these cards on to sheets followed by Xerox reduction on to paper plates for printing 'Recent additions' bulletins on an offset-litho machine.

Costs

Often there is still a choice between different processes and also between different equipment for the same process, and it is at this point that cost must be considered (that is if it hasn't been an overriding factor all the time). The cost of the equipment is important because of the availability or otherwise of the purchase money, but the running costs are even more important, because to a large extent they are concealed costs. Apart from this the Copyright Act 1956 requires those libraries which are permitted to supply photocopies, to charge cost price (including an appropriate allowance for overhead expenses), so it is as well to know with a fair degree of accuracy just how much a copy does cost, otherwise there could be (a) an infringement and (b) a hidden charge against the library's funds which is in effect subsidising the photocopy service.

Because processes and their costs are changing all the time, a straight-forward tabulation of today's costs would be of little use within a short time. It is therefore recommended that full costings be made at the time of selection (and repeated every few years after installation of the equipment). The procedure is as follows:-

1. *Calculate the fixed costs*

(a) DEPRECIATION

Machinery of any kind will ultimately wear out, and the process may become obsolete. For document reproduction equipment it is usually reckoned that printing and duplicating machines will need replacement after 10 years, and photocopying equipment after 8 years. Therefore the annual depreciation of the former is $\frac{1}{10}$th of the capital cost, and of the latter $\frac{1}{8}$th.

(b) INTEREST ON CAPITAL

If the money had not been spent on the purchase of the equipment, it could have been invested, and interest would have accrued. This cost is estimated at 6% per annum.

(c) FLOOR SPACE

The equipment will occupy a measurable amount of space and will require further space round it so that it can be operated in safety. The cost of the space will vary depending on location (e.g. Central London as against say Aberystwyth). As a rough figure use 25*s*. per square foot per annum.

(d) MACHINE LOADING

Although the machine is in the library's possession, it can only be used when the staff are present and when it isn't being repaired or going through a maintenance check. Also there will be periods when there is not sufficient work to keep the machine in full operation. The machine loading will vary considerably, but for library purposes can be estimated at 50%, which will give approximately 1,000 hours machine operation time per annum.

2. *Calculate the variable costs*

(a) COST OF MAINTENANCE AND REPAIRS

In the absence of known cost such as a maintenance contract, this can be estimated at 5% per annum.

(b) MATERIALS COST

The cost of the paper, ink, chemicals, etc., required to make one copy.

(c) LABOUR

The cost of the staff directly involved in the operation of the photocopying machine. This will be related directly to the time taken by the operation.

(d) INDIRECT ONCOST

This covers the cost of heating, lighting, supply of electricity to the machine, supervision, cost of staff over and above wages, etc., and can be estimated at 100% of the labour cost.

3. *The time involved in making copies*

This is not the time which the machine takes to make the copies. In photocopying it is the total time required in taking the originals up from one side of the machine, placing in contact with the light source, switching on, removing the original, developing the copy, and removing it from the machine. It could be argued that the costs of getting the original to the copier and of returning it to the shelves should be included, but in all fairness a similar cost would have been incurred if the person requesting the copy had borrowed the original instead. Such a cost can therefore be considered as a legitimate charge to the normal costs of library service and not specifically charged to the copying service.

In duplicating and offset printing the total time includes the making of the master, the setting up of the machine, the placing of paper into the feeding mechanism and the removal of the printed sheets.

4. *Sample costing*

For an example of the calculation of cost per copy, let us take an imaginary photocopying machine costing £200 which will produce a readable positive at one exposure.

DIRECT ONCOST

Depreciation: ⅛th of capital cost	=	£25.	0.	0
Capital: 6% interest charge	=	£12.	0.	0
Maintenance and repairs: estimated at 5%	=	£10.	0.	0
Floor space: 8 sq. ft. at 25s.	=	£10.	0.	0
		£57.	0.	0

Estimated machine loading 50% (1000 hrs.) = 13.68*d*. per hour = .28*d*. per min.

MATERIALS

Sensitised paper. Cost per packet of 100 sheets	=	£1. 5. 0.
Cost per sheet	=	3.00*d.*
Developing fluid. Estimated cost per copy	=	.05*d.*
Materials cost per copy	=	3.05*d.*

LABOUR

Averaged to £655 per annum = 1.46*d.* per minute.

TIME

Average time taken to place sensitised paper in contact with original, select exposure time, press switch, remove original, feed paper through developing fluid, collect finished copy and stack = 2.5 minutes.

TOTAL COST

Labour: 2.5 minutes at 1.46*d.* per minute	=	3.65*d.*
Materials:	=	3.05*d.*
Direct oncost: 2.5 mins. at .28*d.* per minute	=	.70*d.*
Indirect oncost: at 100% Labour cost	=	3.65*d.*
Total		11.05*d.*

Now it is almost certain that if you were interested in purchasing this machine you would be told by the manufacturer that the cost per copy was about 3*d.* This figure, quoted as a 'materials only' cost is accurate enough, but the true cost of each copy is shewn to be approximately 11*d.* The bulk of this cost is taken up by labour and by indirect oncosts which are directly related to the labour costs. The costs can therefore be kept down by using processes in which the total time occupied in making a copy is less than a minute. However such savings here could be more than offset if the capital or rental charges for the equipment were high.

Location and Layout

Space is at a premium in a library but this should not be used as an excuse to put the document reproduction unit in any old airless cubby-hole. For most of us the old days of dark-rooms are past, and there is no reason why the unit should not be housed in an adequately ventilated and well-lit room, preferably with windows facing north in order to avoid direct sunlight. Processes that require subdued room lighting can be operated away from the windows and with their own light control. Microfilm and microfiche cameras which need to be operated under steady light conditions can be screened or can have dark curtains to cut out variable stray light, and such curtains can be drawn back out of the way when the camera is not in use.

The location and layout of the unit should be principally considered from the point of view of workflow. Two other points to be kept in mind are (a) the noise generated by operating the processes should not be allowed to disturb either readers or staff outside the unit, and (b) the unit should not be located near any source of vibration, because this can affect some of the processes, particularly the cameras, where vibration can result in loss of image definition. For the same reason the cameras should not be located too near to copying and duplicating machinery.

The workflow should be examined in its totality. The material to be copied starts its journey to the unit from various places. The majority probably originates at service counters so the unit should be located as close as possible to the counter which provides the heaviest work-load, or at a strategic point which is no great distance from any counter. This becomes even more important if a 'while-you-wait' service is provided for urgently required copies. The distance which material has to be taken and returned in order to be copied should be kept to the absolute minimum. Two points need to be mentioned here. Where micro-copying heavy bound material, or material which is likely to suffer by being moved about too much, it may often be best to take the cameras to the originals rather than vice-versa. The other point is that if you have a press-button dry copying machine, it may be preferable to have it on or near the service counter in the library and let the readers make their own copies. If this is done, safeguards and controls will have to be built in to the system, but in terms of customer-satisfaction and saving of staff time it may well be worth while.

Inside the unit, like processes should be grouped together and there should be ample working space round the machines. A set of shelves or pigeonholes should be designated for incoming material and should be divided up into spaces, each of which is allocated to particular processes, with an additional space into which urgent material is placed so that it gets priority treatment. A further set should be designated for outgoing material, and this should be divided into spaces allocated to the departments to which the material is to be returned. The different processes will have to be planned separately because each will be dealing with different types of material or with different departments, but each system devised should dove-tail into the others in order to provide full control and as smooth a work flow as possible.

Production cannot be planned in advance because the work-load is dependent, on any particular day, on such variable factors as the number of readers requiring photocopies, and the number of new books for which catalogue cards are required. However plans can be made that will reduce over-loading and also keep down total costs. The most economical way of operating a machine is to get continuous use out of it, but, as we are unlikely to do this in libraries, we must go for the next best which is continuous use on a batch basis. As far as possible all the work for one machine during one day should be done as one batch, thus keeping the machine in use until the work is done, when it can be switched off and left. This will save a lot of to-and-fro movement by the staff operating the machines. If the service routine makes it difficult to operate on the one batch per day basis, then one batch in the morning and one in the afternoon can be acceptable if the work-load is high enough. Alternatively if there is a low work-load then a batch every second day may be the most economical way of operating, although obviously there will be a fall off in the service level.

Forms and Control

There are two kinds of control which are of concern to the librarian:- (a) control of copying in order to ensure that an infringement of the copyright regulations does not take place (the person applying for the photocopy is required to make a declaration, for full details see Chapter 6), and (b) control of all types of document reproduction to ensure that the right process is used; that the work is carried out with the minimum of delay; the quality of the work is of a good order; and that the work is correctly despatched and (where necessary) charged.

Correct design of the forms used will ensure that the control takes place. Theoretically one form should be sufficient for all document reproduction work, but it may be necessary to have one for copies being made for customers, and a separate one for work being done for the library itself.

The information required on the form(s) is as follows:-

1. Title, page numbers, etc., of item to be copied.
2. Process required (photocopy, microfilm, microfiche, etc.)
3. Any degree of urgency.
4. Name and address of customer.
5. Form of declaration.
6. Charge (if any).
7. Receipt.
8. Department or Library dealing with request.
   (Or, in the case of work being done for the Library itself, Department or Library originating the request.)
9. Time request received.
10. Time originals despatched to document reproduction unit.
11. Time work completed.
12. Checked for correct process, quality, etc. and initialled by person i/c unit.
13. Time work despatched for mail or return to department or Library.

The form can be completed in triplicate by the person requiring the copy. One copy is retained by the reader, as a receipt, one copy is retained as a record by the counter and one copy travels with the work. The three copies should have different colours in order to distinguish them. This last copy has the times of action etc., entered on it, so that priorities can be seen easily, and the person in charge of the unit can check that there is no unnecessary hold up on any process.

Charges should be made at a flat rate per page, and a list of charges should be published so that a reader can calculate easily the total cost to him of a photocopy. In order to save invoicing procedure, the charges should, as far as possible, be pre-paid. This will not, of course, be practicable for orders received by post. The page charge should be based on the full cost to the library, not just the materials cost; plus the cost of sending by post. If a while-you-wait service is operated for urgent requests, then the page cost should be higher to cover the disturbance factor, and also the uneconomical use of the machines. This increase is difficult to ascertain but should be put at about 50% of the base rate, because this should be sufficient to discourage people who could easily wait two days (1 for making the copy, 1 for time in the post) for the copies from insisting on immediate copies. Thus if we are using a process where our costing figures show a total of 11*d.* per page, a fair charge would be 1*s.* per page for normal two-day service (including packing and postage), and 1*s.* 6*d.* per page for the while-you-wait service. The costs of packaging can be kept down by the use of standard sizes of envelopes to hold copies made on standard sizes of paper.

Staff

In the majority of libraries, there will not at present be sufficient demand for photocopies to justify the employment of staff to do this work and nothing else. The work has therefore to be combined with other jobs and in the interests of economy, it should be done at as low a level as possible. A photocopy made by a senior member of the professional staff is a very expensive photocopy indeed! Training in the use of the machines will usually be given by the company which supplied the equipment. The person in charge of the unit should be the first to be trained and can then teach the junior staff who are to do the routine operations. Correct training is essential if the costs of the service are to be kept to the minimum. A high wastage of material due to errors in copying can add appreciably to the total cost.

Those libraries which have a large enough demand to set up a separately staffed unit to deal with document reproduction will be concerned with specialised training in this work. The head of the unit may well be a photographer, with qualifications in the 'reprographic' field, and his staff would be encouraged to take courses in reprography, and possibly study for the qualifications of the Institute of Reprographic Technology.

These qualifications are awarded in one or other of two ways. The first is by studentship, for which the academic requirement is the possession of G.C.E. 'O' level with passes in three subjects including English Language. On entry to the Institute students must be at least eighteen years of age and employed in the reprographic industry. They may then train on part-time day release courses for the intermediate and final examinations. Completion of the final examination leads to the award of Associate Membership of the Institute. Alternatively a person over the age of twenty-one may become an Associate of the Institute and may prepare for a qualifying examination leading to Associate Membership. The part-time courses are at present available at the Coventry College of Art and the Barking Regional College of Technology, and courses will be arranged in other regions as sufficient demand arises.

Staff selected for the D.R. unit should have an aptitude for the work. They will need good eyesight, an interest in technical processes and sufficient artistic ability to be able to produce good quality copies. A further attribute, which may be more difficult to find, is a passion for cleanliness and tidiness. In order to function efficiently the unit must have a surfeit of both, in particular the equipment must be kept clean. This brings us to a very important point that is easily overlooked. Where chemicals such as developing fluids and vapours and cleaning fluids are being used, great care must be taken not to employ people who are sensitised to them. This is difficult to ascertain in advance, but the examining Medical Officer should be specially asked to enquire whether there is a history of dermatitis. If a person working in the unit does develop a sensitivity to any of the chemicals used there, then he or she must be taken off the work immediately and employed elsewhere.

Apart from this problem, it must be remembered that although most of the fluids used are harmless, some of them may be toxic, or likely to catch fire. Hazards specific to particular processes should be queried with the supplier when the process is being considered for purchase. In general, the following precautions should be taken:-

1. No smoking, electric fires, etc. near fluids which could catch fire. Suitable fire extinguishers (preferably CO_2 type because these can be also used safely on fires caused by electrical faults) should always be ready to hand in the unit.

2. Fluids should only be used in well ventilated rooms, and containers should be kept in good condition, tightly closed, and correctly labelled.

3. Staff should avoid spilling fluids and breathing vapours, and should protect their hands with barrier creams or rubber gloves, and their clothes with overalls.

Glossary

AMMONIA PROCESS

A dyeline process in which the diazo and coupling compounds are both on the surface of the sensitised paper. After exposure to light the remaining compounds are developed by ammonia vapour.

APERTURE

The opening which controls the amount of light passing through a lens. In ordinary cameras this opening can usually be varied over a considerable range in order to obtain shorter exposures with the wider apertures or greater depth of focus with the smaller apertures. Microform cameras often have a fixed aperture which is designed to give a reasonable depth of focus in order to accommodate thick books.

APERTURE CARD

A card, usually the size of a standard punched card, which has in it a hole into which a microtransparency, usually a 35 mm. full-frame image, can be inserted.

ARCHIVAL PERMANENCE

Copies which are made and stored under controlled conditions in order to last for an indefinitely long period are said to have archival permanence. (British Standard 1153:1955 covers processing and storage of microfilm.)
See also Commercial permanence.

AUTOPOSITIVE PROCESS

See Direct positive process.

AZO DYE

In the dyeline or diazo process the final image is developed when there is a reaction between the remaining diazo and coupling compounds to form an azo dye.

BOOKHOLDER

An ancillary device which can be used when making micro-copies of thick books. In its base are two spring-loaded platens, one for each side of the book, and they press the book upwards against the glass cover and thus hold it in a position which permits the easy copying of the two adjacent pages.

CAMERA

A photographic apparatus in which the light sensitive material is held at the correct distance from the lens, and is protected from exposure to any light other than that which comes through the lens. Incorporated into it there are such devices as a shutter which when opened allows light to pass through the lens, an aperture diaphragm which controls the amount of light, and winding-on mechanism to move the sensitive material when it has been exposed and replace it with a fresh piece.

See also Flat-bed camera, Rotary camera, Step-and-repeat camera

CARD-TO-CARD PRINTER

A machine which will make duplicate copies of aperture cards.

CHEMICAL TRANSFER PROCESS

A photocopying process in which the negative paper is first exposed using the reflex method and is then inserted into a developing machine simultaneously with a sheet of positive paper. After passing through the developing fluid the two are pressed firmly together by rollers and the image is chemically transferred from the negative to the positive. With some types of negative paper several positive prints can be made from one negative. Prints can be made on to translucent paper or film for use as dyeline masters; or on to plates for use in offset-litho printing. Also known as the Diffusion transfer process.

COMMERCIAL PERMANENCE

Copies which are made and stored under conditions which will allow them to last up to a maximum of 25 years are said to have commercial permanence. (British Standard 1153:1955 covers processing and storage of microfilm).

See also Archival permanence.

CONTACT-COPYING

See Non-camera copying.

See also Direct copying.

CONTINUOUS TONE

An illustration, e.g. a photograph, in which the tones range from black to white with a large number of gradations.

See also Half tone.

COUPLER

In dyeline copying the coupler is a chemical compound which at the development stage reacts with the diazo compound to form an azo dye.

DARKROOM

A room from which all daylight can be excluded and which can be illuminated as required by only safe-lighting, and in which light sensitive materials can be safely handled for loading, unloading and development. For most library copying purposes its use is no longer necessary.

DEVELOPER

A chemical or physical agent which produces a visible image on exposed light sensitive material.

DIAZO PROCESS

See Dyeline process.

DIFFUSION TRANSFER PROCESS

See Chemical transfer process.

DIRECT COPYING

When the original document is translucent and has text on one side only, a photocopy can be made by passing light through the original on to the light sensitive paper when the two are in close contact. This is known as direct copying. It is sometimes incorrectly called contact copying.

See also Reflex copying.

DIRECT POSITIVE PROCESS

In this process a special type of silver halide paper is used which gives a positive image after development and thus avoids the need for an intermediate negative when single-sided translucent originals are being copied. With opaque and/or double-sided originals a reversed or mirror positive is first made and then used to make normal positives. This process is also known as the Autopositive process, and the Self-positive process.

DOCUMENT PAPER

See Silver halide paper.

DRY PROCESS

Any photocopying process in which the development is achieved without dampening or wetting the light sensitive material.

DUO

A microfilming technique, used by certain rotary cameras, in which images are photographed on to half the width of the film, and then the film is reloaded and the next series of images are photographed on to the other half.

DUPLEX

A microfilming technique, used by certain rotary cameras, in which both sides of a document are photographed simultaneously and the images placed side by side on the microfilm.

DYE TRANSFER PROCESS

A duplicating process in which the master is prepared by using an aniline dye containing material. The matter to be duplicated is transferred from the master to the copies until the dye is used up and the copies become too weak to be legible. The transfer can be effected via a moistened gelatine surface, or directly from the master by using a solvent. This latter process is also known as Spirit duplication.

DYELINE PROCESS

A piece of diazo-compound coated paper is placed behind a translucent original and exposed to light. This bleaches away the diazo-compounds except where they are protected by the text of the original. The paper is then passed through a fluid containing coupling compounds which join with the diazo-compounds to form an azo dye copy of the original text. Alternatively the two types of compound can be in the coating of the paper and can be developed by ammonia vapour or by heat. This process is also called the Diazo process.

ELECTROLYTIC PROCESS

The sensitised material in this process consists of a zinc oxide coating on to a metal foil. The high electrical sensitivity of the zinc oxide is lowered when exposed to light. After exposure to light projected through a microfilm negative, the sensitised material is given a brief contact with an electroplating solution while at the same time a direct current is applied between the solution and the metal foil. This causes a metallic deposition from the solution on to the zinc oxide, forming a visible image.

ELECTROSTATIC PROCESS

In this process an image of the original document is projected on to an electrostatically charged plate or drum. The charge leaks away in all areas where light strikes and is thus retained only in the text areas. Small particles of a thermosetting resin are cascaded over the plate or drum and adhere to those areas where the electrostatic charge is still retained. A piece of paper which has been given an electrostatic charge of opposite polarity is then brought into contact with the plate or drum. The resin particles are attracted by the charge, transfer to the paper, and are then fixed into position by heat. The copy can be transferred instead to an offset paper plate and used for offset-litho printing. The process is also known as the Xerographic process.

An alternative version of this process uses a special coated copy paper instead of the plate or drum. This allows the copy to be made directly on to copy paper.

ENLARGEMENT

An enlarged print usually a positive print made from a negative film.

ENLARGER

Photographic apparatus used to make enlargements.

EXPOSING UNIT

See Light box.

EXPOSURE

The exposing of a light sensitive material to light, the amount of light reaching the material being controlled by the lens aperture and by the amount of light which is applied to the original.

FACSIMILE

An exact copy of an original.

FACSIMILE TRANSMISSION

A process for line scanning an original and sending the information in the form of impulses by telephone wire or radio wave to a receiving instrument where an exact copy of the original is produced.

FILM

A transparent plastic base coated with a light sensitive emulsion.

FIXING SOLUTION

A solution (hypo) which is used to remove the undeveloped silver salts from light sensitive materials. This 'fixes' the image so that the material can then be handled in normal light. After fixing, the material must be washed in order to remove the residual hypo, otherwise it has a deleterious effect and reduces the permanence of the copy.

FLASH CARD

A card with special markings which is photographed at the appropriate stages when documents are being microfilmed, in order to provide easy visual access when the microfilm is later being run through a microfilm reader.

FLAT-BED CAMERA

A microfilm camera which is usually supported on a vertical column, and the document to be photographed is laid flat on a board at the base of the column. Also called a Planetary camera.

FLOW CAMERA

See Rotary camera.

HALF-TONE

Illustrations prepared by photographing the original continuous tone picture through a line screen which breaks the picture up into a series of dots. The dots are of equal intensity but differ in size. Where they are printed close together they represent a dark tone, and where printed well apart, a light tone.

HAND VIEWER

A small magnifier used for consulting microforms. Some models use incident light, and others have a built-in light source.

HARD COPY

An enlarged copy. This term is mainly used for enlarged copies on paper made via a microform reader/printer.

HECTOGRAPHIC PROCESS

See Dye transfer process.

HYPO TEST

A test used to check the amount of residual hypo left in a film after it has been fixed and washed.

INFRA-RED PROCESS

See Thermographic process.

LEADER

A length of film, at the beginning of a roll, used for threading the film into a camera, or microfilm reader.

LENSLESS COPYING

See Non-camera copying.

LIGHTBOX

In its simplest form it is a box, with an electric light inside and a ground glass screen on the top. Most models have a lid, with a pressure pad, which is brought down on top of the original document and the sensitised paper to ensure the closest possible contact between them while the exposure is being made.

LIGHT SENSITIVE

Used to describe materials in which some change takes place when they are exposed to light.

LITHO

See Offset lithography.

MECHANICAL TRANSFER PROCESS

The negative (or matrix) is exposed by the reflex method and placed in the developing fluid. It is then brought firmly into contact with the positive paper and some of the dye which forms the image is mechanically transferred. Prints can be made on to translucent paper and used as master for diazo copies; or on to plates for offset-litho printing. Also known as the Soft gelatine process.

MICRO-CARD

An opaque card carrying a series of microcopies. The term *Microcard* is a trade name which refers exclusively to 5 in. × 3 in. cards bearing microcopies arranged in a specific manner.

MICRO-CARD READER

See Reader.

MICROCOPYING

Copying by means of an optical device which produces a greatly reduced image of the document being copied.

MICROFICHE

A sheet of film containing a series of microcopies. Also known as Microsheet and Microfilm sheet.

MICROFICHE READER

See Reader.

MICROFILM

A roll or strip of film containing a series of microcopies.

MICROFILM INSERT

A strip of microfilm cut for insertion into a microfilm jacket.

MICROFILM JACKET

An envelope made of transparent plastic and designed to hold strips of microfilm.

MICROFILM READER

See Reader.

MICROFILM SHEET

See Microfiche.

MICROFORM

A general term used for all types of microcopies.

MICROFORM READER

See Reader.

MICRO-OPAQUE

Microcopies on a paper or card base.

MICROPHOTOGRAPHY
The production by photographic processes of copies too small to be read with the naked eye.

'MICROPRINT'
A trade name used by Readex Microprint Corporation for micro-opaques produced by a special printing process.

MICROREPRODUCTION
See Microcopying.

MICROSHEET
See Microfiche.

MICRO-TRANSPARENCY
Microcopies on a film base.

MIRROR POSITIVE
See Direct positive process.

NEGATIVE
A copy in which the tones are reversed, i.e. the black text on a white background in the original appear as white text on a black background in the negative. In a true negative the text is laterally reversed as well. Some processes, e.g. *Photostat* use a prism in front of the lens to reverse the image and thus produce a readable negative.

NEGATIVE/POSITIVE PROCESS
A process in which the first exposure produces a negative print which is then used to make, by a second exposure, a positive print.

NON-CAMERA COPYING
A photocopying technique in which the use of a lens is avoided by placing the document to be copied in close contact with the sensitized paper and then exposing to light.

OFFSET-LITHOGRAPHY
A printing process in which a metal, paper or plastic plate is first prepared by typing or drawing on the surface. The matter can be also put on to the plate by photographic means. A liquid repellent is then applied to the plate after which it is fastened to a cylinder in an offset printing machine. As the cylinder rotates

ink and water are fed on to the plate. The ink remains on the image but is prevented by the water from sticking to the non-image areas. The image is then printed on to a rubber roller, which causes it to be laterally reversed, and from the roller it is printed the right way round on to a sheet of paper.

ORIGINAL

The document from which copies are made.

PHOTOCOPYING

The making of copies by a photographic process. In a strict sense this can refer to the making of microcopies as well as ordinary paper prints, but it has come to be used in the narrower sense of applying only to copies which can be read without the aid of an optical device.

'PHOTOSTAT'

A trade name used by Photostat Ltd. for a process in which photocopies are made by using a large camera fitted with a prism in front of the lens. The use of this term as an alternative to 'photocopy' is common but incorrect.

PLANETARY CAMERA

See Flat-bed camera.

POSITIVE

A copy in which the text is readable and the tones are the same as in the original.

PRINT

A copy, usually on paper.

PROJECTION PRINTING

Making enlarged copies by projecting a micro-image through an optical device.

See also Reader/printers.

PSEUDO-PHOTOGRAPHIC PROCESSES

Those photocopying processes which use some other form of energy as well as, or instead of, light.

READER

An optical device in which a micro-image is enlarged by projection and thus made readable with the naked eye. The enlarged image may be projected on to an opaque surface, e.g. a wall, which is exterior to the reader; on to an opaque screen in the base of the reader; or on to a translucent screen on the face of the reader. The lighting and optical system varies according to whether a microtransparency or a micro-opaque is being enlarged. In the former, i.e. microfilm and microfiche, the light is transmitted through the microtransparency and then through the lens. In the latter, i.e. micro-cards and microprint, the light is reflected from the micro-opaque and then transmitted through the lens.

READER/PRINTER

A microform reader which has, either built-in or as an integral attachment, a system for projecting enlarged images on to light-sensitive paper plus the equipment for giving correct exposure and development in order to produce enlarged copies on paper.

READING HEAD

The assembly, on a microform reader, which contains the light source, the lens, and the attachments for holding the microform in the correct position for it to be read. Also used to describe the device on a tape typewriter through which punched paper tape is passed in order to activate the typewriter.

REDUCTION RATIO

The ratio between an original document and its microcopy. It refers to the measurements made from corner to corner diagonally, and is stated in ratio form, e.g. 18:1.

REFLECTOGRAPHY

See Reflex copying.

REFLEX COPYING

When the original document is opaque or has text on both sides, a photocopy can be made by passing light through the sensitised paper on to the original while the two are held in close contact. The light reflects back from the original and forms a negative image on the sensitised paper. After development, etc. the negative is used to print positive copies. Also known as Reflectography.

See also Direct copying.

REPROGRAPHY

A term which is used to mean the whole field of document reproduction.

REVERSED POSITIVE

See Direct positive process.

ROLL MICROFILM

A roll of film containing a series of microcopies.

ROLL MICRO-OPAQUES

A roll of paper containing a series of microcopies.

ROTARY CAMERA

A microfilm camera in which the documents to be copied are fed on to a moving belt which carries them past the lens where they are automatically photographed.

SCREEN

(1) A surface, which may be opaque or translucent, on to which a micro-image is projected in an enlarged form so that it can be easily read.

(2) Opaque fitting, on some microform readers, which shades the reading screen from unwanted exterior light.

SELF-POSITIVE PROCESS

See Direct positive process.

SEMI-DRY

A term used to describe those processes in which the exposed materials are developed by rapid passing through the developing fluid rather than by immersion for a period. The paper base thus does not get soaked and the resulting print dries out in room temperature within a few minutes.

SHELF-LIFE

The length of time during which sensitised materials can be stored without deteriorating to the point at which they become incapable of making good prints.

SILVER HALIDE PAPER

This is normal photocopying paper which turns black when developed after being exposed to light, and which, depending on its sensitivity, requires handling in subdued or safe light. Also known as Document paper.

See also Negative/positive process, Direct positive process.

SOFT GELATINE TRANSFER PROCESS

See Mechanical transfer process

SPIRIT DUPLICATION

A delightful term which conjures up visions of seances but

See Dye transfer process.

STABILIZER

A fluid which can take the place of both the developing and fixing fluids in the normal photographic process. It allows quick development and semi-dry prints, but does not give permanence.

STENCIL PROCESS

A duplication process in which the master is first prepared by cutting a stencil, usually on a typewriter. The master is then fastened to the duplicating machine and ink is fed through the stencil on to the sheets of semi-absorbent duplicating paper.

STRIP MICROFILM

A strip of film containing a short series of microcopies and kept flat.

See also Microfilm insert, Microfilm jacket, Unitised microcopies

STRIP MICRO-OPAQUES

A strip of paper containing a short series of microcopies and kept flat.

See also Unitised microcopies.

TARGET

Any frame in a microcopy which is devoted to the provision of information for bibliographical control.

THERMOGRAPHIC PROCESS

A pseudo-photographic process in which the original is first exposed to infra-red light which is absorbed by the text and then radiates in the form of heat. Heat sensitive paper is then used to make a copy of the original.

TRAILER

A length of film at the end of a roll which remains in the spool and thus facilitates re-winding of the film.

TRANSFER PROCESS

A photocopying process in which the image is transferred from the negative to the positive while both are held in close contact.

See also Chemical transfer process, Mechanical transfer process.

TRANSLUCENT SCREEN

A sheet of ground glass or of plastic material, fitted to the face of a microform reader. The enlarged image is projected on to the back of the screen and read through it.

UNITISED MICROCOPIES

A series of microcopies, usually dealing with one subject, prepared at different times but kept together in a holder, or on a base. E.g. microfilm can be cut up into short strips and filed subject-wise in microfilm jackets, or cut up into frames and inserted into aperture cards; and roll micro-opaques can similarly be cut up and stuck on to subject cards.

WAX-STENCIL PROCESS

See Stencil process.

XEROGRAPHY

See Electrostatic process.

Sources for further information

Document reproduction is a field in which technology does not stand still. New processes and new or modified equipment appear at such frequent intervals that even the enthusiast finds it a problem to keep abreast of the subject. The object of this short section is to indicate some of the sources which report new developments, and which supply information on the practical applications of processes and equipment. However there is no real substitute for actually seeing equipment in operation and this should be done wherever possible. A visit to the Repro Exhibitions in London, or to one of the Business Efficiency Exhibitions held in London and in the provinces will provide the opportunity to compare and contrast a whole range of equipment. Trade literature, obtained from the manufacturers, will give useful details and specifications, and demonstrations can usually be seen at the manufacturers or retailers showrooms.

The Microfilm Association of Great Britain was established in 1961 (as the Council for Microphotography and Document Reproduction) for the promotion of the use of reprographic techniques, materials and equipment; and to serve user, service and manufacturing interests and needs in the fields of microfilm and document reproduction. It has four groups to cover all aspects of its membership. The groups are as follows:- Group A, Library users (including Universities, Learned Societies, and Archival Associations); Group B, Business and industrial users; Group C, Service organisations and microtext publishers; Group D, Manufacturers of microfilm equipment and material. The Association and its

Groups issue news-letters but their principal publication is *Microdoc*, a journal usually issued quarterly. Each issue contains specialist papers, notes and news, details of new equipment and new microtexts, and a few abstracts of papers published in other journals.

The National Microfilm Association of America has for its purposes the promotion of the lawful interests of the micro-reproduction industry in the direction of good business ethics; the liberal discussion of subjects pertaining to the industry; technological improvement and research; standardisation; methods of manufacturing and marketing; and the education of the consumer in the use of microfilm and related techniques. The official journal is entitled *The National Micro-news* and is published six times a year. Issues contain specialist papers, reports of committees, e.g. committees dealing with microform standards, news and technical notes. The NMA publish the proceedings of their annual conventions, also a *Guide to Microreproduction Equipment*, which is kept up to date between editions by the issue of supplements. Future plans include the replacement of *The National Micro-news* by two separate periodicals, one a quarterly journal and the other a newsletter.

With the assistance of a grant from the Office of Scientific and Technical Information, a National Reprographic Centre for Documentation has been set up in the Hatfield College of Technology, with the County Technical Librarian as Director. The Centre will be particularly concerned with the photographic reduction of documents to microform for storage, handling, retrieval and enlargement; and will act as a national clearing house for information; test and evaluate equipment; examine and evaluate users' needs; and specify areas in which research and development should be carried out. The establishment of this National Centre at Hatfield is recognition of the good work done in this field for many years by the County Technical Librarian, Mr. G. H. Wright, and his colleagues. The papers and discussions of a series of symposia held at Hatfield from 1958–1959 are useful for background reading. The titles are *Microtext and microrecording* (1958); *Modern copying techniques for the drawing office* (1959); and *Microtexts as media for publication* (1959).

The Library Technology Project is a part of the American Library Association, and is supported by annual grants from the Council on Library Resources. Its objects are to test and evaluate library equipment, supplies and systems; to develop new and improved items of library equipment; to promote improvements in existing equipment; to develop performance standards; and to provide an information service to the library profession in order that the results of LTP research may be utilised as effectively as possible. The principal work in the document reproduction field has been done by W. R. Hawken. The published results are as follows:- *Photocopying from bound volumes: a study of machines, methods and materials* (LTP Publication No. 4 and supplements); *Enlarged prints from library microforms: a study of processes, equipment and materials* (LTP Publication No. 6); *Catalog card reproduction* (LTP Publication No. 9); and *Copying methods manual* (LTP Publication No. 11).

The Microfiche Foundation, Delft, The Netherlands, which has already been mentioned in Chapter 3, was founded in 1954 to promote the use of the flat, transparent microfilm sheet in research and industry. It does this through lectures, articles, exhibitions, newsletters, etc., and it stimulates research in the fields of, among others, automation of microfiche cameras, cheap and simple microfiche readers and reader/printers. The Microfiche Foundation *Newsletter* is published about four times a year. There is also a series of publications on the microfiche of which the following are already issued:- No. 1 *Studying with microfiche*; No. 2 *Microfiche for the research worker*; No. 3 *Various applications of the microfiche*; No. 4 *Bibliography on microfiche*; No. 5 *Microfiche equipment.* Publications scheduled for the near future are No. 6 *The Microfiche Foundation* and No. 7 *Microfiche as a medium for publishing material with a limited interest.*

Periodicals which provide access to information in this field include *Library Science Abstracts* (published by the Library Association) which includes a section on document reproduction; *Special Libraries* (published by the Special Libraries Association) which has annually a bibliography on document reproduction compiled by L. J. Kiersky; *Library Resources and Technical Services* (published by the American Library Association) which contains many articles on library aspects of document reproduction together with an annual review of recent developments in the field; *F.I.D. News Bulletin* (published by the Federation Internationale de la Documentation) which contains regular reviews of new developments. Longer term reviews can also be found in the volumes of the Library Association publications *Five years work in librarianship*.

In preparing this textbook the policy has been to omit lists of further reading at the end of each chapter and to make the work reasonably self-contained. There are however some books and periodical articles that will repay consultation and a few are listed here for those students who wish to read more widely in this subject.

1. VERRY, H. R. *Document copying and reproduction processes*. London, Fountain Press, 1958.
2. VERRY, H. R. *Microcopying methods*. London, Focal Press, 1963.
3. LEWIS, C. M. *and* OFFENHAUSER, W. H. *Microrecording: industrial and library applications*. New York, Interscience, 1956.
4. NELSON, C. E. *Microfilm technology: engineering and related fields*. New York, McGraw-Hill, 1965
5. LUTHER, F. *Microfilm: a history*, 1839–1900. Annapolis, National Microfilm Association, 1959.
6. AMERICAN LIBRARY ASSOCIATION, Library Standards for Microfilm Committee. *Microfilm norms: recommended standards for libraries*. Chicago, ALA Resources and Technical Services Division, 1966.
7. SIMONTON, W. "The bibliographical control of microforms" *Library Resources and Technical Services* 6 : Winter 1962, 29–40.
8. GUNTHER, A. "Microphotography in the library". *Unesco Bulletin for Libraries* 16 : January–February 1962, 1–22
9. DUYVIS, F. DONKER. "Document reproduction services : their efficient organization and management". *Unesco Bulletin for Libraries* 14 : November–December 1960, 241–259.

INDEX

39/-